Cambridge Tracts in Mathematics and Mathematical Physics

No. 49

INTEGRAL EQUATIONS

INTEGRAL EQUATIONS

BY

F. SMITHIES

PH.D.

Fellow of St John's College, Cambridge
and Reader in Functional Analysis in the
University of Cambridge

CAMBRIDGE

AT THE UNIVERSITY PRESS

1970

CAMBRIDGE UNIVERSITY PRESS
Cambridge, New York, Melbourne, Madrid, Cape Town, Singapore, São Paulo, Delhi

Cambridge University Press
The Edinburgh Building, Cambridge CB2 8RU, UK

Published in the United States of America by Cambridge University Press, New York

www.cambridge.org
Information on this title: www.cambridge.org/9780521065023

First published 1958
Reprinted 1962, 1965, 1970
This digitally printed version 2008

A catalogue record for this publication is available from the British Library

ISBN 978-0-521-06502-3 hardback
ISBN 978-0-521-10003-8 paperback

CONTENTS

Chapter IV. ORTHONORMAL SYSTEMS OF FUNCTIONS

Chapter V. THE CLASSICAL FREDHOLM THEORY

Chapter VI. THE FREDHOLM FORMULAE FOR $\mathfrak{L}^2$ KERNELS

Chapter VII. HERMITIAN KERNELS

PREFACE

The present work is intended as a successor to Maxime Bôcher's tract *An introduction to the study of integral equations*, which has long been out of print. It is devoted entirely to non-singular linear integral equations, that is, those for which the main results of the Fredholm theory are valid. Only a brief indication of the important applications to differential equations is given, in § 1·2 of the Introduction.

The theory is not presented in terms of linear operators in Hilbert space or a more general topological vector space; this has enabled me to obtain stronger results on the convergence of the various expansions than would have been possible in a more general context. On the other hand, I have made extensive use of the notations of operator theory; many of the formulae thus become much briefer and easier to read.

The Lebesgue integral is used throughout. Most of the theory is given for the case of $\mathfrak{L}^2$ kernels (defined in § 1·6), which illustrate most of the phenomena likely to be encountered; the definition is framed in such a way as to make equations hold everywhere instead of almost everywhere, whenever it is possible to do so. Much information about more general kernels is to be found in A. C. Zaanen's *Linear Analysis* (1953). I have made systematic use of E. H. Moore's relatively uniform convergence (§ 2·4), which seems to fit into the theory in a remarkably natural way.

The literature of the subject is now enormous, and I make no claim to completeness for the historical remarks or for the bibliography; I have given references for the classical results and some of the less familiar ones, and for results not belonging to the theory itself. References in the text to the bibliography are given thus: Fredholm (1903), Schmidt (1907*a*).

My thanks are due to the late G. H. Hardy, who inspired my

early work in the field and first suggested the present enterprise, to my wife for her constant encouragement and for help in preparing the index, to Mr A. L. Brown for assistance in correcting the proofs, and to the Cambridge University Press for the care and accuracy of their printing.

F. S.

ST JOHN'S COLLEGE
CAMBRIDGE
July 1958

NOTE ON THE SECOND IMPRESSION

Almost all the changes in this reprint are corrections of minor errors. My thanks are due to those readers who have made suggestions for corrections and improvements, especially to Professor I. A. Barnett, who provided a most detailed and careful list.

Two excellent books containing much material beyond the scope of this tract have appeared since the bibliography was prepared. These are:

S. G. MIKHLIN (1957). *Integral equations and their applications to certain problems in mechanics, mathematical physics and technology*. Translated by A. H. Armstrong. London, New York, Paris, Los Angeles.

F. G. TRICOMI (1957). *Integral equations*. New York, London.

January 1962

CHAPTER I

INTRODUCTION

1·1. Definitions and examples. An equation in which an unknown function appears under one or more signs of integration is called an *integral equation*. For example, the equations

$$y(s)=\int_a^b K(s,t)\,x(t)\,dt \quad (a\leqslant s\leqslant b), \tag{1}$$

$$x(s)=y(s)+\int_a^b K(s,t)\,x(t)\,dt \quad (a\leqslant s\leqslant b), \tag{2}$$

$$x(s)=\int_a^b K(s,t)\,[x(t)]^2\,dt \quad (a\leqslant s\leqslant b), \tag{3}$$

$$x(s)=\int_a^b\int_a^b K(s,t,u)\,x(t)\,x(u)\,dt\,du \quad (a\leqslant s\leqslant b), \tag{4}$$

in each of which $x(s)$ is the unknown function, and all other functions are regarded as given, are integral equations.

Equations (1) and (2) can be written in the form

$$L[x(s)]=y(s), \tag{5}$$

where the expression $L[x(s)]$ containing the unknown function $x(s)$ is *linear* in the sense that

$$L[\lambda_1 x_1(s)+\lambda_2 x_2(s)]=\lambda_1 L[x_1(s)]+\lambda_2 L[x_2(s)]$$

for any constants λ_1 and λ_2. Thus, for equation (1),

$$L[x(s)]=\int_a^b K(s,t)\,x(t)\,dt,$$

and for (2), $$L[x(s)]=x(s)-\int_a^b K(s,t)\,x(t)\,dt.$$

Equations of the type (5), where $L[x(s)]$ is a linear expression in which $x(s)$ appears under one or more signs of integration, are called *linear integral equations*; thus (1) and (2) are linear, whereas (3) and (4) are not. In this book we shall only be concerned with linear equations.

Linear integral equations of the form (1), where the unknown function appears under the sign of integration and nowhere else in the equation, are called equations of the *first kind*. Those of the form (2), where the unknown function appears both under the sign of integration and elsewhere in the equation, are said to be of the *second kind*. In both (1) and (2), the function $K(s,t)$ is called the *kernel*† of the equation; it is defined in the square $a \leqslant s \leqslant b$, $a \leqslant t \leqslant b$ of the (s,t) plane. The equation is completely specified by giving the interval (a,b), the kernel $K(s,t)$ and the function $y(s)$. In this book we shall be mainly concerned with equations of the second kind, which are the most amenable to general theoretical treatment, but we shall have something to say about equations of the first kind in Chapter VIII.

If we take $y(s)=0$ in (2), we obtain the *homogeneous* equation of the second kind

$$x(s)=\int_a^b K(s,t)\,x(t)\,dt \quad (a \leqslant s \leqslant b). \tag{6}$$

When (2) and (6) are considered simultaneously, (6) is called the homogeneous equation *associated* with (2).

Since an integral equation is usually required to hold for all (or almost all) values of the variable in the range of integration, we shall frequently omit the '$(a \leqslant s \leqslant b)$' on the right-hand side of the equation.

It is often convenient to introduce a parameter λ into equations of the second kind, which then assume the form

$$x(s)=y(s)+\lambda\int_a^b K(s,t)\,x(t)\,dt \quad (a \leqslant s \leqslant b).$$

We can then obtain useful information by studying what happens when λ is allowed to vary in the complex plane.

If $K(s,t)=0$ when $s<t$, equation (2) can be written

$$x(s)=y(s)+\int_a^s K(s,t)\,x(t)\,dt \quad (a \leqslant s \leqslant b),$$

with a variable upper limit of integration. Such an equation is called a *Volterra equation* of the second kind. Similarly,

$$y(s)=\int_a^s K(s,t)\,x(t)\,dt \quad (a \leqslant s \leqslant b)$$

is a Volterra equation of the first kind.

† Called 'noyau' in French 'nucleo' in Italian and 'Kern' in German.

The general equations (1) and (2) are sometimes called *Fredholm equations* of the first and second kinds respectively.

1·2. Connexion with differential equations. The theories of ordinary and partial differential equations are a fruitful source of integral equations. We shall sketch here one of the ways in which integral equations can arise from ordinary differential equations.

We begin by considering the first-order differential equation

$$\frac{dy}{dx} = y' = f(x, y), \tag{1}$$

with the initial condition $y(0) = y_0$. If, say, $f(x, y)$ is a continuous function of (x, y), we can integrate (1) from 0 to x, obtaining

$$y(x) = y_0 + \int_0^x f[t, y(t)]\, dt, \tag{2}$$

an integral equation, in general non-linear, for the function $y(x)$. Conversely, any solution $y(x)$ of (2) clearly satisfies both (1) and the initial condition $y(0) = y_0$. This illustrates the general fact that, by going over to integral equations, we can include both the differential equation and the initial conditions in a single equation.

Let us now consider the second-order differential equation

$$\frac{d^2y}{dx^2} = y'' = f(x, y), \tag{3}$$

with the initial conditions $y(0) = y_0$, $y'(0) = y_1$. We then have

$$y'(x) = y_1 + \int_0^x f[u, y(u)]\, du,$$

whence a second integration gives

$$\begin{aligned} y(x) &= y_0 + y_1 x + \int_0^x dt \int_0^t f[u, y(u)]\, du \\ &= y_0 + y_1 x + \int_0^x f[u, y(u)]\, du \int_u^x dt \\ &= y_0 + y_1 x + \int_0^x (x-u) f[u, y(u)]\, du. \qquad (4) \end{aligned}$$

The argument is reversible, so that here again the differential equation (3), together with the initial conditions, is equivalent to the single integral equation (4).

We see also that any solution of (3) satisfies an integral equation of the form

$$y(x) = A + Bx + \int_0^x (x-u) f[u, y(u)]\, du, \tag{5}$$

the constants A and B being determined by the initial conditions. They may also be determined in other ways; suppose, for instance, that $y(x)$ is required to satisfy a two-point boundary condition, say $y(0)=\alpha$, $y(l)=\beta$. Substituting in (5), we obtain

$$\alpha = y(0) = A,$$

$$\beta = y(l) = A + Bl + \int_0^l (l-u) f[u, y(u)]\, du.$$

Hence $$A=\alpha, \quad B = \frac{\beta-\alpha}{l} - \frac{1}{l}\int_0^l (l-u) f[u, y(u)]\, du.$$

The function $y(x)$ must therefore satisfy the integral equation

$$y(x) = \alpha + \frac{\beta-\alpha}{l}x + \int_0^x (x-u) f[u, y(u)]\, du - \frac{x}{l}\int_0^l (l-u) f[u, y(u)]\, du,$$

which can be written in the form

$$y(x) = z(x) - \int_0^l K(x,u) f[u, y(u)]\, du, \tag{6}$$

where $$z(x) = \alpha + \frac{\beta-\alpha}{l}x,$$

and $$\left.\begin{aligned} K(x,u) &= \frac{u(l-x)}{l} \quad (0 \leqslant u \leqslant x), \\ &= \frac{x(l-u)}{l} \quad (x \leqslant u \leqslant l). \end{aligned}\right\}$$

The argument is again reversible, so that (6) is equivalent to (3) together with the boundary conditions.

If the differential equation is linear, we are led in this way to a linear integral equation of the second kind. For instance, let us start with the linear differential equation

$$y'' + p(x)\, y = \omega(x),$$

which arises in the theory of vibrating strings. Here

$$f[u, y(u)] = \omega(u) - p(u)\, y(u),$$

so that the corresponding integral equation is

$$y(x) = z(x) - \int_0^l K(x, u)\, [\omega(u) - p(u)\, y(u)]\, du$$

$$= w(x) + \int_0^l K(x, u)\, p(u)\, y(u)\, du, \tag{7}$$

where
$$w(x) = z(x) - \int_0^l K(x, u)\, \omega(u)\, du,$$

which is a known function. We see at once that (7) is a linear integral equation of the second kind. Specializing still further, let us consider the equation

$$y'' + \lambda p(x)\, y = 0,$$

with the boundary conditions $y(0) = y(l) = 0$; this arises in the problem of finding the normal modes of vibration of a string with fixed end-points. We are, of course, interested in finding solutions other than the trivial one $y(x) = 0$. The corresponding integral equation is

$$y(x) = \lambda \int_0^l K(x, u)\, p(u)\, y(u)\, du,$$

a homogeneous linear equation of the second kind.

We shall see in Chapter VII that symmetric kernels, i.e. kernels $K(s, t)$ such that $K(s, t) = K(t, s)$, have important special properties. The kernel $K(x, u)\, p(u)$ of (7) is not itself symmetric, but it is easily seen from its definition that $K(x, u)$ is symmetric. If the function $p(x)$ is positive, as it usually is, we can use this fact to transform (7) into an equation with a symmetric kernel; we write it in the form

$$\sqrt{[p(x)]}\, y(x) = \sqrt{[p(x)]}\, w(x)$$
$$+ \int_0^l \sqrt{[p(x)]}\, K(x, u)\, \sqrt{[p(u)]}\, \sqrt{[p(u)]}\, y(u)\, du,$$

i.e.
$$v(x) = g(x) + \int_0^l L(x, u)\, v(u)\, du,$$

where $v(x) = \sqrt{[p(x)]}\, y(x)$, $g(x)$ is a known function, and

$$L(x, u) = \sqrt{[p(x)]}\, K(x, u)\, \sqrt{[p(u)]},$$

which is clearly symmetric. Symmetric kernels often arise in very much this kind of way in problems of mathematical physics.

We shall not have sufficient space to discuss in this book the detailed consequences of this connexion between differential and integral equations; for a full account of this, and for a discussion of the applications of integral equations in potential theory, the reader is referred to Lovitt (1924) and Courant & Hilbert (1953).

1·3. Continuous functions and $\mathfrak{L}^2$ functions. We shall consider throughout complex-valued functions $x(t)$ of a real variable t, defined in a finite interval $a \leqslant t \leqslant b$. Two classes of functions will mainly concern us: continuous functions and functions of integrable square. A function $x(t)$ belongs to the latter class if it is measurable in the interval (a, b) and

$$\int_a^b |x(t)|^2 dt < \infty,$$

the integral being taken in the sense of Lebesgue. Such a function will also be called an $\mathfrak{L}^2$ function.

Since the interval (a, b) will usually be fixed throughout the discussion, we shall often omit the limits of integration.

The restriction to a finite interval (a, b) of the real line is adopted only for reasons of convenience; in the $\mathfrak{L}^2$ case almost everything we shall have to say will be valid, with inessential alterations, for functions defined on any set on which a Lebesgue measure exists, e.g. on an infinite interval, on a set of positive n-dimensional measure in a Euclidean space R^n of n dimensions, or on the surface of a sphere; in the continuous case it may be necessary to restrict the domain of definition to a compact set of finite measure.

If two $\mathfrak{L}^2$ functions $x(t)$ and $y(t)$ are equal for 'almost all' values of t, i.e. except for a set of values of t of Lebesgue measure zero, we shall say that $x(t)$ and $y(t)$ are *equivalent*, and write

$$x(t) =^\circ y(t).$$

If $x(t) =^\circ 0$, we shall call $x(t)$ a *null function*. We shall also use notations such as $x(t) \leqslant^\circ y(t)$ with the obvious meaning. In certain connexions it is customary to regard equivalent functions as being identical; we shall not adopt this convention, but the notion of equivalence will nevertheless play a substantial rôle in our discussions.

The *norm* $\| x \| = \| x \|_c$ of a continuous function $x(t)$ is defined by the equation

$$\| x \|_c = \sup_{a \leqslant t \leqslant b} | x(t) |,$$

the notation 'sup' (for *supremum*) being used to denote the least upper bound. If $x(t)$ is an $\mathfrak{L}^2$ function, we define its $\mathfrak{L}^2$ norm $\| x \| = \| x \|_2$ by the equation

$$\| x \|_2 = \left\{ \int_a^b | x(t) |^2 \, dt \right\}^{\frac{1}{2}}.$$

We note that $\| x \|_c = 0$ if and only if $x(t)$ vanishes identically; on the other hand, $\| x \|_2 = 0$ if and only if $x(t) \overset{\circ}{=} 0$. It will usually be clear from the context which norm is being used.

1·4. The inequalities of Schwarz and Minkowski. We now prove the two fundamental inequalities of the theory of $\mathfrak{L}^2$ functions. The first of these, usually known as *Schwarz's inequality*, is the analogue for integrals of *Cauchy's inequality*† for sequences, which states that if

$$\sum_{n=1}^{\infty} | a_n |^2 < \infty, \quad \sum_{n=1}^{\infty} | b_n |^2 < \infty,$$

then $\sum_{n=1}^{\infty} a_n b_n$ is absolutely convergent, and

$$\left| \sum_{n=1}^{\infty} a_n b_n \right| \leqslant \left(\sum_{n=1}^{\infty} | a_n |^2 \right)^{\frac{1}{2}} \left(\sum_{n=1}^{\infty} | b_n |^2 \right)^{\frac{1}{2}}. \tag{1}$$

THEOREM 1·4·1. *If $x(t)$ and $y(t)$ are $\mathfrak{L}^2$ functions, then $x(t)\, y(t)$ is integrable, and*

$$\left| \int x(t)\, y(t)\, dt \right| \leqslant \| x \| . \| y \|. \tag{2}$$

The function $x(t)\, y(t)$ is clearly measurable. Since

$$\left| \int x(t)\, y(t)\, dt \right| \leqslant \int | x(t) | . | y(t) | \, dt,$$

† Hardy, Littlewood and Pólya (1934), pp. 16, 115.

the existence of the integral on the right implying the existence of that on the left, it is sufficient to prove the result when $x(t)$ and $y(t)$ are real and non-negative. The integrability of $x(t)\,y(t)$ then follows at once from the elementary inequality

$$x(t)\,y(t) \leqslant \tfrac{1}{2}[x(t)]^2 + \tfrac{1}{2}[y(t)]^2.$$

For real λ and μ, the expression

$$\int [\lambda x(t) + \mu y(t)]^2\,dt = \lambda^2 \int [x(t)]^2\,dt + 2\lambda\mu \int x(t)\,y(t)\,dt + \mu^2 \int [y(t)]^2\,dt$$
$$= \alpha\lambda^2 + 2\beta\lambda\mu + \gamma\mu^2,$$

say, is a non-negative definite quadratic form in (λ, μ), so that we must have $\beta^2 \leqslant \alpha\gamma$, which is just the required inequality.

If $x(t)$ and $y(t)$ are $\mathfrak{L}^2$ functions, we define their *inner product*† (x, y) by the equation

$$(x, y) = \int x(t)\,\overline{y(t)}\,dt,$$

where the bar denotes the complex conjugate, as usual. By the theorem we have just proved, (x, y) always exists, and

$$|\,(x, y)\,| \leqslant \|\,x\,\| \,.\, \|\,y\,\|.$$

We also have $\quad (y, x) = \overline{(x, y)}, \quad \|\,x\,\|^2 = (x, x).$

If $(x, y) = 0$, we say that x and y are *orthogonal* to one another; since $(x, y) = 0$ implies $(y, x) = 0$, the relation of orthogonality is symmetrical.

Our second theorem is a particular case of *Minkowski's inequality*.‡

THEOREM 1·4·2. *If $x(t)$ and $y(t)$ are $\mathfrak{L}^2$ functions, then $x(t) + y(t)$ is an $\mathfrak{L}^2$ function, and*

$$\|\,x + y\,\| \leqslant \|\,x\,\| + \|\,y\,\|. \tag{3}$$

It is again sufficient to prove the result when $x(t)$ and $y(t)$ are real and non-negative. If we square both sides of (3), we obtain in this case

$$\|\,x\,\|^2 + 2\int x(t)\,y(t)\,dt + \|\,y\,\|^2 \leqslant \|\,x\,\|^2 + 2\|\,x\,\| \,.\, \|\,y\,\| + \|\,y\,\|^2. \tag{4}$$

† The name 'scalar product' is also used, but tends to be confused with the product of a function by a scalar.

‡ For the general result, see Burkill (1951), p. 66.

Since (4) is an immediate consequence of (2), the result follows at once.

Theorems 1·4·1 and 1·4·2 hold without alteration for functions of more than one variable, the integration being taken over some domain in Euclidean space of the appropriate number of dimensions (and even more generally, in fact); we shall frequently require the two-dimensional case.

For continuous functions, the results corresponding to (2) and (3) are the trivial inequalities

$$\left|\int x(t)\,y(t)\,dt\right| \leqslant (b-a)\,\|x\|_c\,\|y\|_c, \quad \|x+y\|_c \leqslant \|x\|_c + \|y\|_c.$$

It follows from Theorem 1·4·2 that if $x(t)$ and $y(t)$ are $\mathfrak{L}^2$ functions, any linear combination $\lambda x(t)+\mu y(t)$ with constant coefficients is also an $\mathfrak{L}^2$ function. The set of all $\mathfrak{L}^2$ functions therefore forms a complex vector space; the same is clearly true of the set of all continuous functions.

The inner product (x, y) is linear in the first factor and 'anti-linear' in the second, i.e.

$$(\lambda_1 x_1 + \lambda_2 x_2, y) = \lambda_1(x_1, y) + \lambda_2(x_2, y),$$

$$(x, \lambda_1 y_1 + \lambda_2 y_2) = \bar{\lambda}_1(x, y_1) + \bar{\lambda}_2(x, y_2).$$

1·5. Continuous kernels. Let $K(s,t)$ be a continuous function of (s,t) in the square Δ defined by $a \leqslant s \leqslant b$, $a \leqslant t \leqslant b$. If $x(t)$ is a continuous function of t, the function

$$y(s) = \int_a^b K(s,t)\,x(t)\,dt \tag{1}$$

is continuous in $a \leqslant s \leqslant b$. For, given a positive number ϵ, there is a positive number δ, independent of t, such that

$$|K(s,t) - K(s',t)| < \epsilon \quad (|s-s'| < \delta),$$

and, for some $M > 0$, $|x(t)| \leqslant M$ for all t: hence

$$|y(s)-y(s')| \leqslant \int_a^b |K(s,t)-K(s',t)|\,.\,|x(t)|\,dt$$

$$\leqslant \epsilon M(b-a) \quad (|s-s'| < \delta).$$

Equation (1) may be regarded as defining an operator that takes an arbitrary continuous function $x(t)$ into a continuous function $y(s)$, and we shall often write it in the abbreviated form

$$y = Kx. \tag{2}$$

The function $K(s,t)$ is called a *continuous kernel*. If we write

$$\| K \| = \| K \|_c = (b-a) \sup_{(s,t)\in\Delta} | K(s,t) |,$$

we have

$$\| y \|_c \leqslant \| K \|_c \| x \|_c.$$

The operator K is clearly *linear*, in the sense that

$$K(\lambda_1 x_1 + \lambda_2 x_2) = \lambda_1 K x_1 + \lambda_2 K x_2$$

for arbitrary constants λ_1 and λ_2.

If $H(r,s)$ is also a continuous kernel, defined for $a \leqslant r \leqslant b$, $a \leqslant s \leqslant b$, we can form

$$z = Hy = H(Kx).$$

Then
$$z(r) = \int H(r,s)\, ds \int K(s,t)\, x(t)\, dt$$

$$= \int x(t)\, dt \int H(r,s)\, K(s,t)\, ds = \int L(r,t)\, x(t)\, dt,$$

where
$$L(r,t) = \int H(r,s)\, K(s,t)\, ds,$$

which is clearly a continuous function of (r,t). Thus $H(Kx) = Lx$, where L is a continuous kernel. We write $L = HK$, so that we have

$$(HK)\, x = H(Kx);$$

we may also write $L(s,t) = HK(s,t)$. It is easily seen that this 'operator multiplication' of kernels is associative, i.e.

$$G(HK) = (GH)\, K,$$

and doubly distributive with respect to addition, i.e.,

$$G(H+K) = GH + GK, \quad (G+H)\, K = GK + HK.$$

It is not commutative in general, for we do not as a rule have $HK = KH$. Since

$$| HK(r,t) | \leqslant \int | H(r,s) | \,.\, | K(s,t) |\, ds \leqslant \frac{\| H \|_c \| K \|_c}{b-a},$$

we have

$$\| HK \|_c \leqslant \| H \|_c \| K \|_c. \tag{3}$$

By continued operator multiplication, we can form the *iterated kernels* or *iterates* $K^2(s,t)=KK(s,t)$, $K^3(s,t)=KK^2(s,t)$, and so on, of a given kernel $K(s,t)$. We also write $K^1(s,t)=K(s,t)$ It follows at once from the associative law that

$$K^m K^n = K^{m+n}, \quad (K^m)^n = K^{mn} \quad (m, n \geqslant 1).$$

Also, by (3), $\qquad \| K^n \|_c \leqslant \| K \|_c^n \quad (n \geqslant 1).$

Finally, the kernel $K(s,t)$ is completely determined by the corresponding operator, in the sense that if

$$\int K_1(s,t)\,x(t)\,dt = \int K_2(s,t)\,x(t)\,dt \quad (a \leqslant s \leqslant b)$$

for all continuous functions $x(t)$, then $K_1(s,t)=K_2(s,t)$ for all (s,t). For, if we write $K_1 - K_2 = K$, we have

$$\int K(s,t)\,x(t)\,dt = 0 \quad (a \leqslant s \leqslant b)$$

for every continuous $x(t)$. For any fixed s, take $x(t)=\overline{K(s,t)}$; we then have

$$\int | K(s,t) |^2 \,dt = 0,$$

and, since $K(s,t)$ is continuous in t, this implies that $K(s,t)=0$ for all t. Since s is arbitrary, the result follows.

1·6. $\mathfrak{L}^2$ kernels. The situation in the $\mathfrak{L}^2$ case is broadly similar, but there are some differences of detail. We consider functions $K(s,t)$ satisfying the following three conditions:

(*a*) $K(s,t)$ is a measurable function of (s,t) in the square $a \leqslant s \leqslant b$, $a \leqslant t \leqslant b$, such that

$$\int_a^b \int_a^b | K(s,t) |^2 \,ds\,dt < \infty;$$

(*b*) for each value of s, $K(s,t)$ is a measurable function of t such that

$$\int_a^b | K(s,t) |^2 \,dt < \infty;$$

(*c*) for each value of t, $K(s,t)$ is a measurable function of s such that

$$\int_a^b | K(s,t) |^2 \,ds < \infty.$$

A function $K(s, t)$ satisfying these conditions will be called an $\mathfrak{L}^2$ *kernel*.

In dealing with $\mathfrak{L}^2$ kernels, we shall often have occasion to use Fubini's theorem and its extension by Tonelli and Hobson in order to justify changes in the order of integration in multiple integrals. We state these theorems here for convenience. For proofs, see Burkill (1951), pp. 63–4.

THEOREM 1·6·1 (Fubini). *If* $\iint f(s,t)\,ds\,dt$ *exists as a Lebesgue integral, then* $\int f(s,t)\,dt$ *exists for almost all s and is an integrable function of s, and*

$$\iint f(s,t)\,ds\,dt = \int ds \int f(s,t)\,dt.$$

Similarly,

$$\iint f(s,t)\,ds\,dt = \int dt \int f(s,t)\,ds.$$

THEOREM 1·6·2 (Tonelli and Hobson). *If* $f(s,t)$ *is a measurable function of* (s,t) *and any one of the three integrals*

$$\iint |f(s,t)|\,ds\,dt, \quad \int ds \int |f(s,t)|\,dt, \quad \int dt \int |f(s,t)|\,ds$$

exists, then the integrals

$$\iint f(s,t)\,ds\,dt, \quad \int ds \int f(s,t)\,dt, \quad \int dt \int f(s,t)\,ds$$

all exist and are equal to one another.

If $K(s,t)$ is an $\mathfrak{L}^2$ kernel and $x(t)$ is an $\mathfrak{L}^2$ function, it follows at once from (b) and Theorem 1·4·1 that the function

$$y(s) = \int K(s,t)\,x(t)\,dt \tag{1}$$

is defined for all s in (a, b). It is easily seen to be measurable, and since, by Schwarz's inequality,

$$|y(s)|^2 \leqslant \int |K(s,t)|^2\,dt \int |x(t)|^2\,dt,$$

it follows from Fubini's theorem that $y(s)$ is an $\mathfrak{L}^2$ function of s, and that

$$\int |y(s)|^2 ds \leqslant \iint |K(s,t)|^2 ds\,dt \int |x(t)|^2 dt.$$

The equation (1) therefore defines an operator taking $\mathfrak{L}^2$ functions into $\mathfrak{L}^2$ functions, and we write $y = Kx$. The operator K is obviously linear, and if we write

$$\| K \| = \| K \|_2 = \left\{ \iint |K(s,t)|^2 ds\,dt \right\}^{\frac{1}{2}},$$

we have $$\| y \| \leqslant \| K \| . \| x \|.$$

By Minkowski's inequality (Theorem 1·4·2), together with its extension to functions of two variables, the sum of two $\mathfrak{L}^2$ kernels is an $\mathfrak{L}^2$ kernel. The argument leading to the definition of the operator product of two $\mathfrak{L}^2$ kernels is similar to that in the continuous case. To justify the change in the order of integration, we remark that $|H(r,s)|$ and $|K(s,t)|$ are $\mathfrak{L}^2$ kernels and $|x(t)|$ is an $\mathfrak{L}^2$ function, so that

$$\int |H(r,s)|\, ds \int |K(s,t)| . |x(t)|\, dt$$

exists. Since, for each value of r, $H(r,s)\, K(s,t)\, x(t)$ is a measurable function of (s,t), we can apply the Tonelli–Hobson theorem, which gives

$$\begin{aligned} z(r) &= \int H(r,s)\, ds \int K(s,t)\, x(t)\, dt \\ &= \int x(t)\, dt \int H(r,s)\, K(s,t)\, ds = \int L(r,t)\, x(t)\, dt, \end{aligned}$$

where $$L(r,t) = \int H(r,s)\, K(s,t)\, ds.$$

To see that the product $L = HK$ is again an $\mathfrak{L}^2$ kernel, we remark that, by Schwarz's inequality,

$$|L(r,t)|^2 \leqslant \int |H(r,s)|^2 ds \int |K(u,t)|^2 du.$$

Hence†

$$\iint |L(r,t)|^2\,dr\,dt \leqslant \iint dr\,dt \int |H(r,s)|^2\,ds \int |K(u,t)|^2\,du$$

$$= \iint |H(r,s)|^2\,dr\,ds \iint |K(u,t)|^2\,du\,dt$$

$$= \|H\|^2 . \|K\|^2 < \infty, \qquad (2)$$

so that $L(r,t)$ satisfies condition (a). Also, for each value of r,

$$\int |L(r,t)|^2\,dt \leqslant \int |H(r,s)|^2\,ds \iint |K(u,t)|^2\,du\,dt$$

$$= \|K\|^2 \int |H(r,s)|^2\,ds < \infty,$$

so that (b) holds; similarly for condition (c). By (2), we have $\|HK\| \leqslant \|H\| . \|K\|$. The iterated kernels $K^n(s,t)$ are defined as before, and we have $\|K^n\| \leqslant \|K\|^n$.

If conditions (b) and (c) are dropped, so that $K(s,t)$ is only required to satisfy condition (a), we shall call $K(s,t)$ an $\mathfrak{L}^2$ *kernel in the wide sense*; in this case, equation (1) only defines $y(s)$ for almost all s, and it is usually more convenient to adopt the convention mentioned in § 1·3 that equivalent functions are to be regarded as identical. The detailed discussion of this alternative procedure may be left to the reader.

For later use, we remark that if $K(s,t)$ is an $\mathfrak{L}^2$ kernel in the wide sense, there is an equivalent kernel $K_0(s,t)$ satisfying all three conditions (a), (b), (c). For, by Fubini's theorem, $\int |K(s,t)|^2\,dt$ exists for almost all s and is a measurable function of s, and $\int |K(s,t)|^2\,ds$ exists for almost all t and is a measurable function of t. The set of values of (s,t) such that either s or t is exceptional is therefore of plane measure 0; for such values of (s,t) we take $K_0(s,t)=0$, and for the other values of (s,t) we take $K_0(s,t)=K(s,t)$. Clearly $K_0(s,t) \overset{\circ}{=} K(s,t)$, so that

$$\iint |K_0(s,t)|^2\,ds\,dt = \iint |K(s,t)|^2\,ds\,dt < \infty.$$

† Here again we are tacitly using the Tonelli–Hobson theorem; we shall usually do so without mentioning the fact explicitly.

If s is exceptional, $K_0(s,t)=0$ and $\int |K_0(s,t)|^2 dt=0$; if not, $K_0(s,t)=K(s,t)$ for almost all t and

$$\int |K_0(s,t)|^2 dt=\int |K(s,t)|^2 dt,$$

which is finite. Thus $K_0(s,t)$ satisfies condition (b); similarly for (c).

Finally, if the $\mathfrak{L}^2$ kernels K_1 and K_2 define the same operator, then

$$K_1(s,t) =^\circ K_2(s,t).$$

For, if $K=K_1-K_2$, we have

$$\int K(s,t)\,x(t)\,dt=0 \quad (a\leqslant s\leqslant b)$$

for every $\mathfrak{L}^2$ function $x(t)$. For any fixed s, we may take $x(t)=\overline{K(s,t)}$, whence

$$\int |K(s,t)|^2 dt=0; \tag{3}$$

integrating with respect to s, we obtain

$$\iint |K(s,t)|^2 ds\,dt=0,$$

so that $K(s,t) =^\circ 0$, the required result. We also see from (3) that, for any fixed value of s, $K(s,t)=0$ for almost all t, but we cannot conclude in this case that $K(s,t)$ vanishes identically.

In the following chapters of this book we shall usually give a detailed discussion only in the $\mathfrak{L}^2$ case; we shall sometimes state the corresponding results for the continuous case, but the proofs will be omitted unless they differ in some essential way from those given for the $\mathfrak{L}^2$ case.

CHAPTER II

THE RESOLVENT KERNEL AND THE NEUMANN SERIES

2·1. Regular values and the resolvent kernel. We consider the linear integral equation of the second kind

$$x(s) = y(s) + \lambda \int_a^b K(s,t)\, x(t)\, dt \quad (a \leqslant s \leqslant b), \tag{1}$$

where $K(s,t)$ is an $\mathfrak{L}^2$ kernel and $y(s)$ is an $\mathfrak{L}^2$ function. Our problem is to find all $\mathfrak{L}^2$ functions $x(s)$ that satisfy (1). The introduction of the complex parameter λ enables us to study the properties of the equation simultaneously for all the kernels $\lambda K(s,t)$. In our abbreviated notation (1) becomes

$$x = y + \lambda K x. \tag{2}$$

If we introduce the symbol I for the 'identity operator',† so that $Ix = x$ for every $\mathfrak{L}^2$ function $x(s)$, we can also write (2) in the form

$$(I - \lambda K)\, x = y. \tag{3}$$

If (3) has an $\mathfrak{L}^2$ solution x, it is natural to ask whether x can be expressed in terms of y by a formula similar to (3); in other words, whether we can write

$$x = (I + \lambda H)\, y = y + \lambda H y, \tag{4}$$

where $H(s,t) = H_\lambda(s,t)$ is an $\mathfrak{L}^2$ kernel, depending, of course, on the parameter λ. If such an x satisfies (3), we must have

$$(I - \lambda K)(I + \lambda H)\, y = y = Iy;$$

this suggests that H should satisfy the operator equation

$$I - \lambda K + \lambda H - \lambda^2 K H = I,$$

whence, if $\lambda \neq 0$, we obtain

$$H - K = \lambda K H. \tag{5}$$

† There is no kernel $I(s, t)$ with this property, so I must be regarded simply as a symbol of operation. We have $IK = KI = K$ for every $\mathfrak{L}^2$ kernel K.

On the other hand, if $x=(I+\lambda H)y$ is the unique $\mathfrak{L}^2$ solution of (3), we must have

$$x=(I+\lambda H)y=(I+\lambda H)(I-\lambda K)x,$$

which suggests that H should also satisfy the equation

$$H-K=\lambda HK. \tag{6}$$

We now make some formal definitions. If, for a given value of λ, there is an $\mathfrak{L}^2$ kernel $H_\lambda(s,t)$ satisfying (5) and (6), i.e. such that

$$H_\lambda(s,t)-K(s,t)=\lambda\int K(s,u)\,H_\lambda(u,t)\,du=\lambda\int H_\lambda(s,u)\,K(u,t)\,du,$$

then $H_\lambda(s,t)$ is a *resolvent kernel* or *resolvent* of the kernel $K(s,t)$ for the value λ, and λ is a *regular value* of the kernel $K(s,t)$. We call the equation

$$H_\lambda-K=\lambda KH_\lambda=\lambda H_\lambda K \tag{7}$$

the *resolvent equation* of K. We remark that $\lambda=0$ is always a regular value of K, the corresponding resolvent kernel H_0 being just K itself.

THEOREM 2·1·1. *If, for a given value of λ, a resolvent kernel of the $\mathfrak{L}^2$ kernel K exists, then it is unique.*

Suppose that

$$H_1-K=\lambda H_1K=\lambda KH_1, \tag{8}$$

$$H_2-K=\lambda H_2K=\lambda KH_2, \tag{9}$$

and write $H=H_1-H_2$. Subtracting (9) from (8), we obtain

$$H=\lambda KH, \tag{10}$$

whence

$$H_1H=H_1(\lambda KH)=(\lambda H_1K)H=(H_1-K)H=H_1H-KH,$$

so that $KH=0$, and therefore, by (10), $H=0$, i.e. $H_1=H_2$.

We can therefore speak without ambiguity of *the* resolvent kernel of a given $\mathfrak{L}^2$ kernel for a given regular value of the parameter λ.

THEOREM 2·1·2. *Let λ be a regular value of the $\mathfrak{L}^2$ kernel K, and let H be the corresponding resolvent kernel. If y is a given $\mathfrak{L}^2$ function, the equation*

$$x=y+\lambda Kx \tag{11}$$

has a unique $\mathfrak{L}^2$ solution x given by

$$x = y + \lambda H y. \tag{12}$$

If x is defined by (12), substitution in the right-hand side of (11) gives

$$y + \lambda K(y + \lambda H y) = y + \lambda K y + \lambda^2 K H y = y + \lambda K y + \lambda(H - K) y$$
$$= y + \lambda H y = x,$$

so that x is a solution of (11). Conversely, if x satisfies (11), we have $y = x - \lambda K x$, whence

$$y + \lambda H y = x - \lambda K x + \lambda H x - \lambda^2 H K x$$
$$= x + \lambda(H - K) x - \lambda(H - K) x = x;$$

thus x must be given by (12), and the solution is unique.

The requirement that the solution $x(s)$ be an $\mathfrak{L}^2$ function is indispensable for the truth of the theorem; at the end of § 2·7 we shall give an example of an $\mathfrak{L}^2$ kernel with a regular value λ for which (11) has more than one solution; the additional solutions are not, of course, $\mathfrak{L}^2$ functions.

We now consider the continuous case for a moment. If $K(s, t)$ is a continuous kernel, and we require from the beginning that $H(s, t)$ be continuous, the above argument goes through without change. We can say more, however; a continuous kernel is *a fortiori* an $\mathfrak{L}^2$ kernel, and may thus have an $\mathfrak{L}^2$ resolvent H_λ. In this case, we have

$$H_\lambda = K + \lambda K H_\lambda = K + \lambda K(K + \lambda H_\lambda K)$$
$$= K + \lambda K^2 + \lambda^2 K H_\lambda K. \tag{13}$$

We shall deduce from this equation that $H_\lambda(s, t)$ is a continuous kernel. To do this, we require the following preliminary result, which will be useful again later.

THEOREM 2·1·3. *If $K(s, t)$, $H(s, t)$ and $L(s, t)$ are $\mathfrak{L}^2$ kernels, then*

$$| KHL(s, t) | \leqslant \| H \| \, k(s) \, l(t),$$

where $k(s) = \left\{ \int | K(s, t) |^2 dt \right\}^{\frac{1}{2}}, \quad l(t) = \left\{ \int | L(s, t) |^2 ds \right\}^{\frac{1}{2}}.$

For arbitrary $\mathfrak{L}^2$ functions x and y, we have

$$| (Hx, y) | \leqslant \| Hx \| . \| y \| \leqslant \| H \| . \| x \| . \| y \|. \tag{14}$$

Taking, for fixed s and t, $x(v)=L(v,t)$ and $y(u)=\overline{K(s,u)}$, we have

$$(Hx,y)=\iint K(s,u)\,H(u,v)\,L(v,t)\,du\,dv=KHL(s,t);$$

also $\| x \| = l(t)$ and $\| y \| = k(s)$. Hence, by (14),

$$|\,KHL(s,t)\,| \leqslant \| H \| \, k(s)\, l(t),$$

the required result.

We now return to the proof that the continuity of K implies that of H_λ. Since $K+\lambda K^2$ is clearly continuous, it is sufficient, by (13), to show that $KH_\lambda K$ is continuous. We have

$$\begin{aligned}&KH_\lambda K(s,t)-KH_\lambda K(s',t')\\ &\quad =KH_\lambda K(s,t)-KH_\lambda K(s,t')+KH_\lambda K(s,t')-KH_\lambda K(s',t').\end{aligned}$$

By Theorem 2·1·3,

$$\begin{aligned}&|\,KH_\lambda K(s,t)-KH_\lambda K(s,t')\,|\\ &\qquad \leqslant \| H_\lambda \| \, k(s)\left\{\int |\,K(s,t)-K(s,t')\,|^2\,ds\right\}^{\frac{1}{2}}<\tfrac{1}{2}\epsilon \quad (|\,t-t'\,|<\delta_\epsilon),\end{aligned}$$

since $K(s,t)$ and $k(s)$ are continuous. Similarly

$$|\,KH_\lambda K(s,t')-KH_\lambda K(s',t')\,|<\tfrac{1}{2}\epsilon \quad (|\,s-s'\,|<\eta_\epsilon).$$

Hence

$$|\,KH_\lambda K(s,t)-KH_\lambda K(s',t')\,|<\epsilon \quad (|\,s-s'\,|<\eta_\epsilon\ |\,t-t'\,|<\delta_\epsilon),$$

i.e. $KH_\lambda K(s,t)$ is continuous, and the result is proved.

We conclude that every '$\mathfrak{L}^2$ regular value' of a continuous kernel is a 'continuous regular value'; the converse statement is trivial. We can therefore speak of the regular values of a continuous kernel without specifying whether we want a continuous or an $\mathfrak{L}^2$ resolvent; if λ is such a value, the integral equation

$$x(s)=y(s)+\lambda\int K(s,t)\,x(t)\,dt$$

has a unique continuous solution $x(s)$ for any given continuous function $y(s)$.

2·2. The adjoint kernel and the adjoint equation. If $K(s,t)$ is an $\mathfrak{L}^2$ kernel, then $K^*(s,t)=\overline{K(t,s)}$ is also an $\mathfrak{L}^2$ kernel, known as the *Hermitian adjoint* or simply the *adjoint* of $K(s,t)$. The operation of taking the adjoint clearly obeys the rules

$$(K^*)^*=K, \quad \| K^* \| = \| K \|, \quad (\lambda K)^*=\bar{\lambda}K^*, \quad (K+L)^*=K^*+L^*. \tag{1}$$

To find the adjoint of the operator product of two kernels, we remark that

$$(KL)^*(s,t)=\overline{KL(t,s)}=\int \overline{K(t,u)}\,\overline{L(u,s)}\,du$$
$$=\int L^*(s,u)\,K^*(u,t)\,du=L^*K^*(s,t),$$

so that
$$(KL)^*=L^*K^*. \tag{2}$$

If $x(s)$ and $y(s)$ are $\mathfrak{L}^2$ functions, we have

$$(K^*x,y)=\int \overline{y(s)}\,ds\int \overline{K(t,s)}\,x(t)\,dt$$
$$=\int x(t)\,dt\int \overline{K(t,s)}\,\overline{y(s)}\,ds=(x,Ky);$$

by taking the complex conjugate and interchanging x and y, we obtain $(Kx,y)=(x,K^*y)$.

If $K=K^*$, $K(s,t)$ is said to be *Hermitian symmetric* or *Hermitian*, and if $KK^*=K^*K$, $K(s,t)$ is said to be *normal*; we remark that every Hermitian kernel is *a fortiori* normal.

If $K(s,t)$ is real-valued, its adjoint is the same as the *transposed* kernel $\tilde{K}(s,t)=K(t,s)$ of K; in this case K is Hermitian if and only if it is *symmetric*, i.e. $K(s,t)=K(t,s)$.

The integral equation
$$u=v+\bar{\lambda}K^*u \tag{3}$$
is called the *adjoint equation* of
$$x=y+\lambda Kx. \tag{4}$$

THEOREM 2·2·1. *If K is an $\mathfrak{L}^2$ kernel, λ is a regular value of K if and only if $\bar{\lambda}$ is a regular value of K^*. The resolvent of K^* for $\bar{\lambda}$ is $(H_\lambda)^*$.*

Let λ be a regular value of K, and H_λ the corresponding resolvent. Then

$$H_\lambda - K = \lambda H_\lambda K = \lambda K H_\lambda. \tag{5}$$

Taking adjoints in (5), we obtain

$$(H_\lambda)^* - K^* = \bar{\lambda} K^* (H_\lambda)^* = \bar{\lambda} (H_\lambda)^* K^*;$$

thus $(H_\lambda)^*$ satisfies the resolvent equation of K^* for $\bar{\lambda}$, and is therefore the required resolvent. The converse follows at once from the equation $(K^*)^* = K$.

2·3. Characteristic values and characteristic functions. We now turn for a moment to the homogeneous linear integral equation of the second kind

$$x(s) = \lambda \int_a^b K(s,t)\, x(t)\, dt \quad (a \leqslant s \leqslant b), \tag{1}$$

or, in our abbreviated notation,

$$x = \lambda K x. \tag{2}$$

This equation has the trivial solution $x(s) = 0$. If it has any $\mathfrak{L}^2$ solution $x(s)$ other than this trivial one, we shall call λ a *characteristic value* of $K(s,t)$, and the function $x(s)$ a *characteristic function* of $K(s,t)$ belonging to the characteristic value λ. If $x_1(s)$ and $x_2(s)$ are both characteristic functions belonging to λ, so is any non-zero linear combination $\alpha_1 x_1(s) + \alpha_2 x_2(s)$ with constant coefficients. In other words, the characteristic functions belonging to λ, together with the function 0, form a vector subspace of the complex vector space consisting of all $\mathfrak{L}^2$ functions; we call this the *characteristic subspace* belonging to λ.

If $x(s)$ is a characteristic function of $K(s,t)$, then $\| x \| \neq 0$; for, if $\| x \| = 0$, we should have $x(s) \overset{\circ}{=} 0$, whence we see by substitution in the right-hand side of (1) that $x(s) = 0$ for all s, a contradiction.

It is sometimes more convenient to write the homogeneous equation in the form

$$Kx = \kappa x. \tag{3}$$

If (3) has a non-zero $\mathfrak{L}^2$ solution x, we call κ an *eigenvalue*† of K, and x an *eigenfunction* belonging to κ. If λ is a characteristic value of K, then λ^{-1} is an eigenvalue ($\lambda=0$ is clearly never a characteristic value), and the characteristic functions belonging to λ are precisely the eigenfunctions belonging to λ^{-1}. Conversely, if κ is a non-zero eigenvalue of K, then κ^{-1} is a characteristic value. The value $\kappa=0$ is, according to our definition, always an eigenvalue, with the null functions (not identically zero) as eigenfunctions, and it may have other eigenfunctions as well; the eigenfunctions belonging to $\kappa=0$ are sometimes loosely described as characteristic functions belonging to the 'characteristic value' ∞.

THEOREM 2·3·1. *No regular value of an $\mathfrak{L}^2$ kernel K can be a characteristic value of K.*

If λ is a regular value of K, and $x=\lambda Kx$, we have

$$x=y+\lambda Kx \tag{4}$$

with $y=0$. By Theorem 2·1·2, (4) has the unique $\mathfrak{L}^2$ solution

$$x=y+\lambda H_\lambda y=0,$$

so λ cannot be a characteristic value.

We shall prove later (Theorem 3·6·1) that every complex number λ is either a regular value or a characteristic value of a given $\mathfrak{L}^2$ kernel; for the moment our main concern is with the regular values.

Turning to the continuous case, we remark that if $x(s)$ is an $\mathfrak{L}^2$ characteristic function of a continuous kernel $K(s,t)$, it follows at once from the equation

$$x(s)=\lambda\int K(s,t)\,x(t)\,dt$$

that $x(s)$ is continuous; for we have, by Schwarz's inequality,

$$|x(s)-x(s')|\leqslant|\lambda|\,\|x\|\left\{\int|K(s,t)-K(s',t)|^2\,dt\right\}^{\frac{1}{2}}.$$

† The terminology in the literature is very variable. The names 'characteristic value' and 'eigenvalue' are sometimes interchanged, and sometimes no distinction is made between them. The name 'proper value' is also used for one or other of them. 'Eigenvalue' seems to have become established in quantum mechanics with the above meaning.

Thus we need not distinguish in this case between $\mathfrak{L}^2$ and continuous characteristic functions.

On the other hand, a continuous kernel can have a discontinuous $\mathfrak{L}^2$ eigenfunction (other than a null function) belonging to the eigenvalue 0. To illustrate this, we take

$$K(s,t) = 2s - 5st \quad (0 \leqslant s \leqslant 1,\ 0 \leqslant t \leqslant 1)$$

and

$$x(t) = t^{-\frac{1}{3}} \quad (0 < t \leqslant 1),$$

$$= 0 \quad (t = 0).$$

Then a trivial calculation shows that

$$\int_0^1 K(s,t)\, x(t) = 0 \quad (0 \leqslant s \leqslant 1).$$

2·4. Relatively uniform convergence. Before embarking on the discussion of methods of finding the resolvent of a given kernel, we define the notion of relatively uniform convergence.†

A sequence $\{x_n(s)\}$ of $\mathfrak{L}^2$ functions is said to be *relatively uniformly convergent* to the limiting function $x(s)$ if there exists a non-negative $\mathfrak{L}^2$ function $p(s)$ such that, given $\epsilon > 0$, there is a positive integer $n_0(\epsilon)$ for which

$$|x_n(s) - x(s)| \leqslant \epsilon p(s) \quad (n \geqslant n_0,\ a \leqslant s \leqslant b).$$

Since this implies that

$$|x(s)| \leqslant |x_n(s)| + \epsilon p(s) \quad (n \geqslant n_0,\ a \leqslant s \leqslant b),$$

the limit $x(s)$ is an $\mathfrak{L}^2$ function; the sequence $\{x_n(s)\}$ is clearly convergent to $x(s)$ in the ordinary sense for each value of s. An infinite series of $\mathfrak{L}^2$ functions is said to be relatively uniformly convergent if the sequence formed by its partial sums is relatively uniformly convergent. Finally, an infinite series

$$\sum_{n=1}^{\infty} x_n(s)$$

of $\mathfrak{L}^2$ functions is said to be *relatively uniformly absolutely convergent* if the series $\Sigma\, |x_n(s)|$ is relatively uniformly convergent; we note that a series with this property is both relatively uniformly convergent and absolutely convergent.

† Introduced by E. H. Moore (1910), pp. 30ff.

Similarly, a sequence $\{K_n(s,t)\}$ of $\mathfrak{L}^2$ kernels will be said to be relatively uniformly convergent to $K(s,t)$ if there is a non-negative $\mathfrak{L}^2$ kernel $P(s,t)$ such that, given $\epsilon > 0$, there is a positive integer $n_0(\epsilon)$ for which

$$|K_n(s,t) - K(s,t)| \leqslant \epsilon P(s,t) \quad (n \geqslant n_0;\ a \leqslant s \leqslant b,\ a \leqslant t \leqslant b).$$

The limit $K(s,t)$ is again an $\mathfrak{L}^2$ kernel. The only non-trivial point in the proof of this is the measurability of the functions

$$\int |K(s,t)|^2\,dt, \quad \int |K(s,t)|^2\,ds;$$

we see this by remarking that, by Minkowski's inequality,

$$\left|\left\{\int |K_n(s,t)|^2\,dt\right\}^{\frac{1}{2}} - \left\{\int |K(s,t)|^2\,dt\right\}^{\frac{1}{2}}\right| \leqslant \left\{\int |K_n(s,t) - K(s,t)|^2\,dt\right\}^{\frac{1}{2}}$$

$$\leqslant \epsilon\left\{\int |P(s,t)|^2\,dt\right\}^{\frac{1}{2}} \quad (n \geqslant n_0,\ a \leqslant s \leqslant b).$$

Thus

$$\int |K(s,t)|^2\,dt = \lim_{n\to\infty} \int |K_n(s,t)|^2\,dt,$$

and is therefore measurable; similarly for the other expression. Definitions for infinite series can be made as in the case of $\mathfrak{L}^2$ functions.

The following theorem summarizes the principal properties of relatively uniform convergence that we shall need later.

THEOREM 2·4·1. (*a*) *A necessary and sufficient condition for the relatively uniform convergence of a sequence* $\{x_n(s)\}$ *of* $\mathfrak{L}^2$ *functions is that there exist a non-negative* $\mathfrak{L}^2$ *function* $p(s)$ *with the property that, for any* $\epsilon > 0$, *there is a positive integer* $n_0(\epsilon)$ *such that*

$$|x_n(s) - x_m(s)| \leqslant \epsilon p(s) \quad (n \geqslant n_0,\ m \geqslant n_0;\ a \leqslant s \leqslant b). \tag{1}$$

(*b*) *The analogous statement holds for* $\mathfrak{L}^2$ *kernels.*

(*c*) *If* $x_n(s) \to x(s)$ $(n \to \infty)$ *relatively uniformly, and* $y(s)$ *is an* $\mathfrak{L}^2$ *function, then*

$$(x_n, y) \to (x, y) \quad (n \to \infty).$$

(*d*) *If* $K_n(s,t) \to K(s,t)$ $(n \to \infty)$ *as a relatively uniformly convergent sequence of* $\mathfrak{L}^2$ *kernels, and* $x(t)$ *is an* $\mathfrak{L}^2$ *function, then*

$$\int K_n(s,t)\,x(t)\,dt \to \int K(s,t)\,x(t)\,dt \quad (n \to \infty)$$

relatively uniformly.

(e) *If* $K_n(s,t) \to K(s,t)$ $(n \to \infty)$ *relatively uniformly, and* $L(s,t)$ *is an* $\mathfrak{L}^2$ *kernel, then*

$$K_n L(s,t) \to KL(s,t), \quad LK_n(s,t) \to LK(s,t) \quad (n \to \infty),$$

the convergence being relatively uniform in each case.

(f) *If*
$$\sum_{n=1}^{\infty} x_n(s) = x(s),$$

the series being relatively uniformly absolutely convergent, then

$$\sum_{n=1}^{\infty} (x_n, y) = (x, y),$$

the series on the left-hand side being absolutely convergent.

(g) *If*
$$\sum_{n=1}^{\infty} K_n(s,t) = K(s,t),$$

the left-hand side being a relatively uniformly absolutely convergent series of $\mathfrak{L}^2$ *kernels, and* $x(t)$ *is an* $\mathfrak{L}^2$ *function, then*

$$\sum_{n=1}^{\infty} \int K_n(s,t)\, x(t)\, dt = \int K(s,t)\, x(t)\, dt,$$

the series being relatively uniformly absolutely convergent.

(h) *If*
$$\sum_{n=1}^{\infty} K_n(s,t) = K(s,t),$$

the series being relatively uniformly absolutely convergent, and $L(s,t)$ *is an* $\mathfrak{L}^2$ *kernel, then*

$$\sum_{n=1}^{\infty} K_n L(s,t) = KL(s,t), \quad \sum_{n=1}^{\infty} LK_n(s,t) = LK(s,t),$$

the series being relatively uniformly absolutely convergent in each case.

In (*a*), the necessity of the condition is obvious. To prove sufficiency, we need only remark that $\{x_n(s)\}$ is a Cauchy sequence of complex numbers for each value of s, and therefore converges to a limit $x(s)$; letting $m \to \infty$ in (1), we have

$$|x_n(s) - x(s)| \leqslant \epsilon p(s) \quad (n \geqslant n_0,\ a \leqslant s \leqslant b),$$

whence the result.

The proof of (*b*) is similar to that of (*a*). The proofs of (*c*), (*d*) and (*e*) all run along the same lines; we give that of the first part of (*e*). We have

$$|K_nL(s,t)-KL(s,t)| = \left|\int \{K_n(s,u)-K(s,u)\}L(u,t)\,du\right|$$

$$\leqslant \epsilon \int P(s,u)\,|L(u,t)|\,du \quad (n \geqslant n_0);$$

since $\int P(s,u)\,|L(u,t)|\,du$ is an $\mathfrak{L}^2$ kernel, the result follows at once.

The proofs of (*f*), (*g*) and (*h*) are again all on the same lines; to prove the first part of (*h*), for instance, we note first that equality follows from (*e*); secondly, since $\Sigma\,|K_n(s,t)|$ is a relatively uniformly convergent series of $\mathfrak{L}^2$ kernels and $|L(s,t)|$ is an $\mathfrak{L}^2$ kernel, the series $\Sigma \int |K_n(s,u)|\,.\,|L(u,t)|\,du$ is relatively uniformly convergent, by (*e*), and the convergence statement of (*h*) then follows at once from the inequality

$$|K_nL(s,t)| \leqslant \int |K_n(s,u)|\,.\,|L(u,t)|\,du.$$

In defining a corresponding notion for continuous functions, we would naturally take the majorizing function $p(s)$ to be continuous; we then obtain just uniform convergence in the ordinary sense. A similar remark applies to continuous kernels. The statements analogous to those of Theorem 2·4·1 then follow from well-known results on uniform convergence.

2·5. The Neumann series. We have shown that if λ is a regular value of the $\mathfrak{L}^2$ kernel $K(s,t)$, the equation

$$x = y + \lambda Kx \tag{1}$$

has the unique solution

$$x = y + \lambda H_\lambda y, \tag{2}$$

where H_λ is the resolvent of K. In order to solve (1) when λ is a regular value, it is therefore sufficient to find an explicit expression for H_λ. One way of doing this is suggested by the resolvent equation

$$H_\lambda - K = \lambda H_\lambda K = \lambda K H_\lambda,$$

which we write in the form

$$H_\lambda(I-\lambda K) = (I-\lambda K)\,H_\lambda = K.$$

A purely formal argument then gives

$$\begin{aligned} H_\lambda &= K(I-\lambda K)^{-1} \\ &= K(I+\lambda K+\lambda^2K^2+\dots) \\ &= K+\lambda K^2+\lambda^2K^3+\dots, \end{aligned} \tag{3}$$

where we have expanded $(I-\lambda K)^{-1}$ as a geometric series.

Another idea that leads to the series (3) is that of solving (1) by a process of successive approximation. If we begin by neglecting the term λKx, we obtain, as a first approximation for x,

$$x_1 = y.$$

Substituting x_1 for x in the right-hand side of (1), we obtain as the second approximation

$$x_2 = y+\lambda Ky.$$

Putting x_2 for x gives as the third approximation

$$x_3 = y+\lambda Ky+\lambda^2K^2y.$$

Continuing this process, we find as the nth approximation

$$x_n = y+\lambda Ky+\lambda^2K^2y+\dots+\lambda^{n-1}K^{n-1}y.$$

This suggests that the infinite series

$$x = y+\lambda Ky+\lambda^2K^2y+\dots \tag{4}$$

may be a solution of (1). Since (4) may also be written as

$$x = y+\lambda(K+\lambda K^2+\dots)\,y,$$

we have again arrived at the expression (3) for the resolvent H_λ.

We now state a precise theorem corresponding to the above non-rigorous arguments.

THEOREM 2·5·1. *If $K(s,t)$ is an $\mathfrak{L}^2$ kernel, the series*

$$K(s,t)+\lambda K^2(s,t)+\lambda^2K^3(s,t)+\dots \tag{5}$$

is relatively uniformly absolutely convergent for all λ such that $|\lambda|.\|K\|<1$, and its sum $H_\lambda(s,t)$ is the resolvent of $K(s,t)$. Thus every value of λ such that $|\lambda|.\|K\|<1$ is a regular value of $K(s,t)$.

If $n > 2$, we have, by Theorem 2·1·3,

$$\begin{aligned} |\lambda^{n-1}K^n(s,t)| &= |\lambda|^{n-1}|KK^{n-2}K(s,t)| \\ &\leqslant |\lambda|^{n-1}\|K^{n-2}\|\,k_1(s)\,k_2(t) \\ &\leqslant |\lambda|^{n-1}\|K\|^{n-2}\,k_1(s)\,k_2(t), \end{aligned} \qquad (6)$$

where $k_1(s) = \left\{\int |K(s,t)|^2\,dt\right\}^{\frac{1}{2}}, \quad k_2(t) = \left\{\int |K(s,t)|^2\,ds\right\}^{\frac{1}{2}}.$

Since $k_1(s)$ and $k_2(t)$ are $\mathfrak{L}^2$ functions, $k_1(s)\,k_2(t)$ is an $\mathfrak{L}^2$ kernel, and it follows from Theorem 2·4·1 (b) that (5) is relatively uniformly absolutely convergent when $|\lambda|.\|K\| < 1$. If we now write

$$H_\lambda(s,t) = K(s,t) + \lambda K^2(s,t) + \lambda^2 K^3(s,t) + \ldots,$$

we have to show that

$$H_\lambda - K = \lambda H_\lambda K = \lambda K H_\lambda.$$

By Theorem 2·4·1 (h), we have

$$\begin{aligned} \lambda H_\lambda K &= \lambda(K + \lambda K^2 + \lambda^2 K^3 + \ldots)K \\ &= \lambda K^2 + \lambda^2 K^3 + \lambda^3 K^4 + \ldots \\ &= H_\lambda - K; \end{aligned}$$

similarly $\lambda K H_\lambda = H_\lambda - K$, and the proof is complete.

The condition $|\lambda|.\|K\| < 1$ is by no means a necessary one for the convergence of (5). For instance, let

$$K(s,t) = u(s)\,\overline{v(t)},$$

where u and v are $\mathfrak{L}^2$ functions such that $(u,v) = 0$. Then

$$K^2(s,t) = \int u(s)\,\overline{v(r)}\,u(r)\,\overline{v(t)}\,dr = 0;$$

all the higher iterates K^n vanish, and we have $H_\lambda = K$. The series is therefore convergent, in a trivial sense, for all values of λ. A less trivial example in which (5) converges for all λ is furnished by the Volterra kernels discussed in § 2·7.

If $\|K\| = 0$, so that $K(s,t)$ is a null function, the inequality (6) shows that $K^n(s,t) = 0$ for $n > 2$, so that the resolvent H_λ reduces to $K+\lambda K^2$. On the other hand, the kernel $K^2(s,t)$ need not vanish identically; for instance, take

$$K(s,t)=0 \quad (s\neq\tfrac{1}{2},\ t\neq\tfrac{1}{2}),$$
$$=1 \quad (s=\tfrac{1}{2} \text{ or } t=\tfrac{1}{2}),$$

in $0\leqslant s\leqslant 1$, $0\leqslant t\leqslant 1$. Then $K^2(\tfrac{1}{2},\tfrac{1}{2})=1$, although $K^2(s,t)=0$ for all other values of (s,t). We note that $K^2x=0$ for all x; nevertheless the term λK^2 cannot be dropped from the resolvent.

THEOREM 2·5·2. *If $K(s,t)$ is an $\mathfrak{L}^2$ kernel, $y(s)$ is an $\mathfrak{L}^2$ function, and $|\lambda|.\|K\|<1$, the integral equation*

$$x(s)=y(s)+\lambda\int K(s,t)\,x(t)\,dt \tag{7}$$

has the unique $\mathfrak{L}^2$ solution

$$x(s)=y(s)+\lambda y_1(s)+\lambda^2 y_2(s)+\dots, \tag{8}$$

where
$$y_n(s)=\int K^n(s,t)\,y(t)\,dt \quad (n\geqslant 1),$$

the series in (8) *being relatively uniformly absolutely convergent.*

In particular, the homogeneous equation

$$x(s)=\lambda\int K(s,t)\,x(t)\,dt \tag{9}$$

has the unique $\mathfrak{L}^2$ solution $x(s)=0$.

By Theorem 2·5·1, (7) has the unique $\mathfrak{L}^2$ solution

$$x=y+\lambda H_\lambda y, \tag{10}$$

where
$$H_\lambda=K+\lambda K^2+\lambda^2K^3+\dots. \tag{11}$$

By Theorem 2·4·1 (*g*), it follows from (11) that

$$H_\lambda y=Ky+\lambda K^2y+\lambda^2K^3y+\dots,$$

the series being relatively uniformly absolutely convergent. The result now follows by substituting this series in (10).

Theorem 2·5·2 could also be proved directly without appealing to the properties of the resolvent kernel; the details may be left to the reader.

The series (8) is known as the *Liouville–Neumann*† or *Neumann series*. By extension, (5) may be called the Neumann series for the resolvent.

The analogues of Theorems 2·5·1 and 2·5·2 for continuous kernels and functions follow at once from the theorems themselves, relatively uniform convergence being automatically replaced by uniform convergence.

2·6. Generalization of the Neumann series. In the last section we have seen that for an $\mathfrak{L}^2$ kernel $K(s,t)$ all sufficiently small complex numbers λ, in other words, all numbers near enough to the regular value $\lambda=0$, are regular values. We shall now show that if λ_0 is a given regular value, all numbers λ near enough to λ_0 are regular values; the set $\rho(K)$ of regular values of K is therefore an open set in the complex plane.

THEOREM 2·6·1. *If λ_0 is a regular value of the $\mathfrak{L}^2$ kernel $K(s,t)$, and $H_0(s,t)$ is the corresponding resolvent,‡ then every value of λ such that $|\lambda-\lambda_0|.\|H_0\|<1$ is a regular value of $K(s,t)$, and the corresponding resolvent is*

$$H_\lambda = H_0 + (\lambda-\lambda_0)H_0^2 + (\lambda-\lambda_0)^2 H_0^3 + \dots, \tag{1}$$

the series (1) *being relatively uniformly absolutely convergent.*

The convergence of the series is proved as for (5) of § 2·5. By the resolvent equation for $\lambda=\lambda_0$, we have, for $n \geqslant 2$,

$$H_0^n = H_0 H_0^{n-1} = (K+\lambda_0 KH_0)H_0^{n-1} = KH_0^{n-1} + \lambda_0 KH_0^n.$$

Hence
$$\begin{aligned} H_\lambda &= H_0 + \sum_{n=2}^{\infty} (\lambda-\lambda_0)^{n-1} H_0^n \\ &= K + \lambda_0 KH_0 + \sum_{n=2}^{\infty} (\lambda-\lambda_0)^{n-1}(KH_0^{n-1} + \lambda_0 KH_0^n) \\ &= K + K\left\{\sum_{n=1}^{\infty}(\lambda-\lambda_0)^n H_0^n + \sum_{n=1}^{\infty}\lambda_0(\lambda-\lambda_0)^{n-1}H_0^n\right\} \\ &= K + K[(\lambda-\lambda_0)+\lambda_0]\sum_{n=1}^{\infty}(\lambda-\lambda_0)^{n-1}H_0^n \\ &= K + \lambda KH_\lambda, \end{aligned}$$

† J. Liouville (1837); C. Neumann (1877), ch. 5.
‡ Note that H_0 is not here the value of H_λ for $\lambda=0$.

the extraction of the factor K being justified by Theorem 2·4·1 (*e*). Similarly

$$H_\lambda = K + \lambda H_\lambda K.$$

Thus λ is a regular value of K with resolvent H_λ.

A less direct but more algebraic proof begins by remarking that H_λ, as defined by (1), satisfies the equation

$$H_\lambda - H_0 = (\lambda - \lambda_0) H_\lambda H_0 = (\lambda - \lambda_0) H_0 H_\lambda. \tag{2}$$

We also have
$$H_0 - K = \lambda_0 H_0 K = \lambda_0 K H_0. \tag{3}$$

By eliminating H_0 from (2) and (3) we can conclude that H_λ satisfies the resolvent equation; we leave the details as an exercise to the reader. It can also be shown without difficulty that (2) holds for any two regular values λ and λ_0; this equation is sometimes called the *functional equation of the resolvent.*

COROLLARY. *In the open set $\rho(K)$ of regular values of the $\mathfrak{L}^2$ kernel $K(s,t)$, the resolvent kernel $H_\lambda(s,t)$ is, for all (s,t), an analytic function of the complex variable λ.*

For, in the neighbourhood of any point λ_0 of $\rho(K)$, $H_\lambda(s,t)$ is given by a power series in $(\lambda - \lambda_0)$.

There is an analogue of Theorem 2·5·2, but we shall not state it explicitly; we also omit the statement of the result corresponding to Theorem 2·6·1 for continuous kernels.

2·7. Volterra integral equations of the second kind.

We have defined (§ 1·1) a Volterra integral equation of the second kind as an equation of the form

$$x(s) = y(s) + \lambda \int_a^s K(s,t)\, x(t)\, dt \quad (a \leqslant s \leqslant b), \tag{1}$$

the upper limit of integration being the variable s instead of a constant b. If we extend the domain of definition of $K(s,t)$ by defining it to be zero when $a \leqslant s < t \leqslant b$, (1) can be written in our standard form

$$x(s) = y(s) + \lambda \int_a^b K(s,t)\, x(t)\, dt \quad (a \leqslant s \leqslant b). \tag{2}$$

We shall suppose throughout this section that $K(s,t)$ is defined in the square $a \leqslant s \leqslant b$, $a \leqslant t \leqslant b$, vanishes when $a \leqslant s < t \leqslant b$, and is an $\mathfrak{L}^2$ kernel, and that $y(s)$ is an $\mathfrak{L}^2$ function in $a \leqslant s \leqslant b$. A kernel $K(s,t)$ of this type will be called a *Volterra kernel*.

The operator product of two Volterra kernels is again a Volterra kernel. For, let

$$L(s,t) = \int_a^b H(s,u)\,K(u,t)\,du, \tag{3}$$

where H and K are Volterra kernels. Since $H(s,u) = 0$ when $s < u$, and $K(u,t) = 0$ when $u < t$, it follows at once that $L(s,t) = 0$ when $s < t$, i.e. $L(s,t)$ is a Volterra kernel; we remark that equation (3) can therefore be written in the form

$$L(s,t) = \int_t^s H(s,u)\,K(u,t)\,du \quad (a \leqslant t \leqslant s \leqslant b).$$

In particular, all the iterates $K^n(s,t)$ of a Volterra kernel $K(s,t)$ are Volterra kernels.

Our main result will be that the Neumann series for the resolvent of a Volterra kernel converges for all complex λ, so that the set $\rho(K)$ of regular values is the entire plane.† We require two preliminary lemmas.

LEMMA 1. *Let $K(s,t)$ be an $\mathfrak{L}^2$ Volterra kernel and $x(s)$ an $\mathfrak{L}^2$ function. If*

$$x_n(s) = \int_a^s K^n(s,t)\,x(t)\,dt \quad (n = 1, 2, \ldots),$$

then $$|x_n(s)| \leqslant \frac{k_1(s)\,\|x\|}{\sqrt{[(n-1)!]}} \left\{\int_a^s [k_1(t)]^2\,dt\right\}^{\frac{1}{2}(n-1)} \quad (a \leqslant s \leqslant b), \tag{4}$$

where $$k_1(s) = \left\{\int_a^s |K(s,t)|^2\,dt\right\}^{\frac{1}{2}}.$$

When $n = 1$, (4) follows at once from Schwarz's inequality. Supposing that (4) holds for n, we shall deduce it for $n+1$. Since

$$x_{n+1}(s) = \int_a^s K(s,t)\,x_n(t)\,dt,$$

† For continuous kernels, this result was first proved by J. Le Roux (1895), pp. 243ff., and V. Volterra (1897); for $\mathfrak{L}^2$ kernels by F. Smithies (1935); see also F. Riesz (1910), pp. 491–6.

Schwarz's inequality implies that

$$|x_{n+1}(s)|^2 \leqslant \int_a^s |K(s,t)|^2 dt . \int_a^s |x_n(t)|^2 dt = [k_1(s)]^2 \int_a^s |x_n(t)|^2 dt. \tag{5}$$

Applying (4), for the positive integer n, to the right-hand side of (5), we obtain

$$|x_{n+1}(s)|^2 \leqslant [k_1(s)]^2 \frac{\|x\|^2}{(n-1)!} \int_a^s [k_1(t)]^2 dt \left\{\int_a^t [k_1(u)]^2 du\right\}^{n-1}. \tag{6}$$

We now remark that

$$\left\{\int_a^t [k_1(u)]^2 du\right\}^{n-1}$$
$$= \int_a^t [k_1(u_1)]^2 du_1 \int_a^t [k_1(u_2)]^2 du_2 \ldots \int_a^t [k_1(u_{n-1})]^2 du_{n-1},$$

which is an $(n-1)$-fold integral over the region $a \leqslant u_\nu \leqslant t$ $(1 \leqslant \nu \leqslant n-1)$ in R^{n-1}. This region is the union of the $(n-1)!$ regions defined by

$$a \leqslant u_{p_1} \leqslant u_{p_2} \leqslant \ldots \leqslant u_{p_{n-1}} \leqslant t,$$

where $(p_1, p_2, \ldots, p_{n-1})$ runs over all permutations of $(1, 2, \ldots, n-1)$; no two of these regions have interior points in common, and the integral over each of them has the same value. Hence

$$\left\{\int_a^t [k_1(u)]^2 du\right\}^{n-1}$$
$$= (n-1)! \int_a^t [k_1(u_1)]^2 du_1 \int_a^{u_1} [k_1(u_2)]^2 du_2 \ldots \int_a^{u_{n-2}} [k_1(u_{n-1})]^2 du_{n-1}. \tag{7}$$

The inequality (6) therefore becomes

$$|x_{n+1}(s)|^2 \leqslant [k_1(s)]^2 \|x\|^2 \int_a^s [k_1(t)]^2 dt \int_a^t [k_1(u_1)]^2 du_1$$
$$\times \int_a^{u_1} [k_1(u_2)]^2 du_2 \ldots \int_a^{u_{n-2}} [k_1(u_{n-1})]^2 du_{n-1}$$
$$= [k_1(s)]^2 \frac{\|x\|^2}{n!} \left\{\int_a^s [k_1(t)]^2 dt\right\}^n,$$

by a second application of (7), this time for n instead of $n-1$. The inequality (4) therefore holds for $n+1$ if it holds for n, and the lemma is established.

Lemma 2. *If $K(s,t)$ is an $\mathfrak{L}^2$ Volterra kernel, then*

$$|K^{n+1}(s,t)| \leqslant \frac{\|K\|^{n-1}}{\sqrt{[(n-1)!]}} k_1(s)\, k_2(t) \quad (n=1,2,\ldots), \tag{8}$$

where $k_1(s)$ is defined as in Lemma 1, and

$$k_2(t) = \left\{\int_t^b |K(s,t)|^2 ds\right\}^{\frac{1}{2}}.$$

Since $K(s,t)$ is an $\mathfrak{L}^2$ function for all t, we can take $x(s)=K(s,t)$ in Lemma 1; we then have

$$\|x\| = k_2(t), \quad x_n(s) = K^{n+1}(s,t).$$

Also
$$\int_a^s [k_1(t)]^2\, dt \leqslant \int_a^b [k_1(t)]^2\, dt = \|K\|^2.$$

The result then follows at once from the inequality (4).

Theorem 2·7·1. *If $K(s,t)$ is an $\mathfrak{L}^2$ Volterra kernel, the series*

$$K(s,t) + \lambda K^2(s,t) + \lambda^2 K^3(s,t) + \ldots \tag{9}$$

is relatively uniformly absolutely convergent for every complex number λ, and its sum $H_\lambda(s,t)$ is the resolvent of $K(s,t)$.

The convergence statement follows at once from the inequality (8) of Lemma 2, and the remainder of the proof goes exactly as for Theorem 2·5·1.

As a consequence of Theorem 2·7·1, equation (1) has a unique $\mathfrak{L}^2$ solution $x(s)$ for any given $\mathfrak{L}^2$ function $y(s)$ and every complex number λ.

The resolvent $H_\lambda(s,t)$ is a Volterra kernel, and it is an integral function of λ for any given (s,t).

The statement and proof of the analogue of Theorem 2·5·2 may be left to the reader.

Since, by Theorem 2·3·1, no regular value is a characteristic value, the homogeneous equation

$$x(s) = \lambda \int_a^s K(s,t)\, x(t)\, dt \quad (a \leqslant s \leqslant b)$$

has the unique $\mathfrak{L}^2$ solution $x(s)=0$ for every value of λ. The importance of restricting ourselves to $\mathfrak{L}^2$ solutions is illustrated by the following example (Bôcher (1909), p. 17). Take

$$K(s,t)=t^{s-t} \quad (0<t\leqslant s\leqslant 1),$$
$$=0 \quad (t=0);$$

then $K(s,t)$ is bounded and measurable in $0\leqslant t\leqslant s\leqslant 1$, and is therefore an $\mathfrak{L}^2$ Volterra kernel. The homogeneous equation

$$x(s)=\int_0^s t^{s-t}x(t)\,dt \quad (0\leqslant s\leqslant 1)$$

nevertheless has the non-zero solution $x_0(s)$, given by

$$x_0(s)=s^{s-1} \quad (0<s\leqslant 1),$$
$$=0 \quad (s=0).$$

Clearly $x_0(s)$ is not an $\mathfrak{L}^2$ function. Furthermore, if $x(s)$ is any solution of the non-homogeneous equation

$$x(s)=y(s)+\int_0^s t^{s-t}x(t)\,dt \quad (0\leqslant s\leqslant 1),$$

then $x(s)+Ax_0(s)$, where A is an arbitrary constant, is also a solution; nevertheless, there is only one $\mathfrak{L}^2$ solution.

When $K(s,t)$ is continuous in $a\leqslant t\leqslant s\leqslant b$, its resolvent $H_\lambda(s,t)$ is continuous in the same region, since the series (9) is uniformly convergent; incidentally, the proof of its convergence can be made somewhat simpler than that given above for the general $\mathfrak{L}^2$ case. If also $y(s)$ is continuous in $a\leqslant s\leqslant b$, every $\mathfrak{L}^2$ solution of the equation

$$x(s)=y(s)+\lambda\int_a^s K(s,t)\,x(t)\,dt \tag{10}$$

is automatically continuous. Thus (10) has a unique continuous solution for every complex number λ.

CHAPTER III

THE FREDHOLM THEOREMS

3·1. Introductory remarks. In the last chapter we have shown that, for an $\mathfrak{L}^2$ kernel, all sufficiently small values of the parameter λ are regular, and that the regular values form an open set in the complex plane. We shall now discuss a method of finding the general $\mathfrak{L}^2$ solution of the integral equation

$$x(s)=y(s)+\lambda\int K(s,t)\,x(t)\,dt, \tag{1}$$

where $K(s,t)$ is an $\mathfrak{L}^2$ kernel and $y(s)$ is an $\mathfrak{L}^2$ function, and λ may or may not be a regular value of $K(s,t)$. This method was introduced by E. Schmidt (1907*b*), who called it the 'Abspaltungsverfahren' (dissection process), and it was further developed by J. Radon (1919).

We begin with a particularly simple case, that of kernels of finite rank, and we then use the results for this case to construct a method that is applicable to general $\mathfrak{L}^2$ kernels. We do not, however, obtain an explicit formula for the solution of (1) in terms of $K(s,t)$ and $y(s)$ and valid for all regular values of λ; such formulae will be developed in Chapters V and VI.

3·2. Kernels of finite rank. A function $K(s, t)$ that can be expressed as a finite sum of the form

$$K(s,t)=\sum_{\nu=1}^{n} a_\nu(s)\,\overline{b_\nu(t)} \quad (a\leqslant s\leqslant b,\ a\leqslant t\leqslant b), \tag{1}$$

where $a_\nu(s)$ and $b_\nu(t)$ are $\mathfrak{L}^2$ functions $(1\leqslant \nu\leqslant n)$, is clearly an $\mathfrak{L}^2$ kernel. We shall say that any such kernel is of *finite rank*, and we shall write

$$K=\sum_{\nu=1}^{n} a_\nu\otimes b_\nu. \tag{2}$$

If K is given by (2), we have

$$\alpha K=\sum_{\nu=1}^{n}(\alpha a_\nu)\otimes b_\nu=\sum_{\nu=1}^{n} a_\nu\otimes(\bar{\alpha} b_\nu)$$

for any scalar α, and
$$K^* = \sum_{\nu=1}^{n} b_\nu \otimes a_\nu.$$

If x is any $\mathfrak{L}^2$ function,
$$Kx = \sum_{\nu=1}^{n} (x, b_\nu)\, a_\nu,$$

and if H is an $\mathfrak{L}^2$ kernel,

$$HK = \sum_{\nu=1}^{n} (Ha_\nu) \otimes b_\nu, \quad KH = \sum_{\nu=1}^{n} a_\nu \otimes (H^* b_\nu).$$

The smallest integer n for which K can be expressed in the form (2) is called the *rank* of K. If the number of terms in (2) is equal to the rank of K, the functions a_ν are linearly independent over the complex field, for otherwise one of the a_ν could be expressed as a linear combination of the others, and we should be able to express K as a sum of the form (2) with fewer terms; similarly the functions b_ν are linearly independent. Conversely, if both (a_ν) and (b_ν) are linearly independent, the rank of K is exactly n; for, if

$$\sum_{\nu=1}^{n} a_\nu(s)\, \overline{b_\nu(t)} = \sum_{\mu=1}^{m} c_\mu(s)\, \overline{d_\mu(t)},$$

where $m < n$, it follows from the linear independence of (b_ν) that we can find n values t_ρ such that the determinant of the matrix $[b_{\nu\rho}]$, where $b_{\nu\rho} = b_\nu(t_\rho)$, is non-zero; we then have

$$\sum_{\nu=1}^{n} \overline{b_{\nu\rho}}\, a_\nu(s) = \sum_{\mu=1}^{m} \overline{d_\mu(t_\rho)}\, c_\mu(s) \quad (1 \leqslant \rho \leqslant n),$$

a system of equations that can be solved for the functions $a_\nu(s)$; the resulting solution expresses them as linear combinations of the m functions $c_\mu(s)$, contradicting the linear independence of (a_ν).

THEOREM 3·2·1. *Let*

$$K = \sum_{\nu=1}^{n} a_\nu \otimes b_\nu$$

be an $\mathfrak{L}^2$ kernel of finite rank, let y be an $\mathfrak{L}^2$ function, and write $(y, b_\mu) = y_\mu$, $(a_\nu, b_\mu) = k_{\mu\nu}$ $(1 \leqslant \mu \leqslant n,\ 1 \leqslant \nu \leqslant n)$. If x is an $\mathfrak{L}^2$ solution of the integral equation

$$x = y + \lambda Kx, \tag{3}$$

and $$x_\mu = (x, b_\mu) \quad (1 \leqslant \mu \leqslant n),$$

then $$x_\mu = y_\mu + \lambda \sum_{\nu=1}^{n} k_{\mu\nu} x_\nu \quad (1 \leqslant \mu \leqslant n). \tag{4}$$

Conversely, if (x_μ) *is a solution of the system of linear equations* (4), *and*

$$x = y + \lambda \sum_{\nu=1}^{n} x_\nu a_\nu, \tag{5}$$

then x *is an* $\mathfrak{L}^2$ *solution of* (3) *and* $(x, b_\mu) = x_\mu$.

If x satisfies (3), we have

$$x = y + \lambda \sum_{\nu=1}^{n} (x, b_\nu) a_\nu = y + \lambda \sum_{\nu=1}^{n} x_\nu a_\nu;$$

forming the inner product of this equation with b_μ, we obtain (4). Conversely, if x is defined by (5), it is clearly an $\mathfrak{L}^2$ function, and we have

$$\begin{aligned}(x, b_\mu) &= (y, b_\mu) + \lambda \sum_{\nu=1}^{n} x_\nu (a_\nu, b_\mu) \\ &= y_\mu + \lambda \sum_{\nu=1}^{n} k_{\mu\nu} x_\nu \\ &= x_\mu,\end{aligned}$$

by (4), whence (5) becomes

$$x = y + \lambda \sum_{\nu=1}^{n} (x, b_\nu) a_\nu = y + \lambda K x,$$

i.e. x is a solution of (3).

Thus, when the kernel of an integral equation is of finite rank n, the problem of solving the equation is reduced to that of solving a system of n linear equations in n unknowns; there is a (1, 1) correspondence between $\mathfrak{L}^2$ solutions x of (3) and solutions (x_μ) of (4).

If we use matrix notation, writing $\mathbf{x}$ for the column vector $[x_\mu]$, $\mathbf{K}$ for the matrix $[k_{\mu\nu}]$, and so on, (4) becomes

$$\mathbf{x} = \mathbf{y} + \lambda \mathbf{K} \mathbf{x}; \tag{6}$$

we see at once that (6) has a similar structure to the original integral equation.

Let us write (6) in the form

$$(\mathbf{I} - \lambda \mathbf{K}) \mathbf{x} = \mathbf{y}, \tag{7}$$

where $\mathbf{I}=[\delta_{\mu\nu}]$ is the $n\times n$ unit matrix. If the determinant

$$d(\lambda)=\det(\mathbf{I}-\lambda\mathbf{K})$$

does not vanish, the inverse matrix $(\mathbf{I}-\lambda\mathbf{K})^{-1}$ exists; we have in fact

$$(\mathbf{I}-\lambda\mathbf{K})^{-1}=\frac{1}{d(\lambda)}\operatorname{adj}(\mathbf{I}-\lambda\mathbf{K})=\frac{1}{d(\lambda)}\mathbf{D}_\lambda,$$

say, where $\operatorname{adj}(\mathbf{I}-\lambda\mathbf{K})$ denotes the adjugate matrix of $\mathbf{I}-\lambda\mathbf{K}$, i.e. the transposed matrix of cofactors; the determinant $d(\lambda)$ and the elements $d_{\mu\nu}(\lambda)$ of $\mathbf{D}_\lambda$ are polynomials in λ. Equation (7) then has the unique solution

$$\mathbf{x}=\frac{\mathbf{D}_\lambda\mathbf{y}}{d(\lambda)},$$

i.e.

$$x_\mu=\frac{1}{d(\lambda)}\sum_{\nu=1}^{n}d_{\mu\nu}(\lambda)\,y_\nu\quad(1\leqslant\mu\leqslant n).\tag{8}$$

Substituting in (5) from (8), we see that the integral equation (3) has the unique solution

$$\begin{aligned}x&=y+\frac{\lambda}{d(\lambda)}\sum_{\mu=1}^{n}\sum_{\nu=1}^{n}d_{\mu\nu}(\lambda)\,(y,b_\nu)\,a_\mu\\&=y+\lambda H_\lambda y,\end{aligned}$$

where

$$H_\lambda=\frac{1}{d(\lambda)}\sum_{\mu,\nu=1}^{n}d_{\mu\nu}(\lambda)\,(a_\mu\otimes b_\nu).$$

These results suggest the following theorem.

THEOREM 3·2·2. *Let*

$$K=\sum_{\nu=1}^{n}a_\nu\otimes b_\nu$$

be an $\mathfrak{L}^2$ *kernel of finite rank, and write*

$$\mathbf{K}=[k_{\mu\nu}]=[(a_\nu,b_\mu)],\quad d(\lambda)=\det(\mathbf{I}-\lambda\mathbf{K}),$$

$$\mathbf{D}_\lambda=\operatorname{adj}(\mathbf{I}-\lambda\mathbf{K})=[d_{\mu\nu}(\lambda)].$$

Then any complex number λ *such that* $d(\lambda)\neq 0$ *is a regular value of* K, *and the corresponding resolvent is given by*

$$H_\lambda=\frac{1}{d(\lambda)}\sum_{\mu,\nu=1}^{n}d_{\mu\nu}(\lambda)\,(a_\mu\otimes b_\nu).$$

We verify directly that H_λ satisfies the resolvent equation; we have

$$\begin{aligned}\lambda K H_\lambda &= \frac{\lambda}{d(\lambda)} \sum_{\mu,\nu=1}^{n} d_{\mu\nu}(\lambda)\,(Ka_\mu \otimes b_\nu) \\ &= \frac{\lambda}{d(\lambda)} \sum_{\rho,\mu,\nu=1}^{n} k_{\rho\mu} d_{\mu\nu}(\lambda)\,(a_\rho \otimes b_\nu) \\ &= \frac{1}{d(\lambda)} \sum_{\rho,\nu=1}^{n} [d_{\rho\nu}(\lambda) - d(\lambda)\,\delta_{\rho\nu}]\,(a_\rho \otimes b_\nu) \\ &= H_\lambda - K,\end{aligned}$$

where we have used the equation $\lambda \mathbf{K}\mathbf{D}_\lambda = \mathbf{D}_\lambda - d(\lambda)\,\mathbf{I}$, which follows at once from the definitions of $d(\lambda)$ and $\mathbf{D}_\lambda$. Similarly

$$\lambda H_\lambda K = H_\lambda - K.$$

Since $d(\lambda)$ is a polynomial, the resolvent H_λ exists except perhaps for a finite number of values of λ, and it is a rational function of λ.

We shall not consider here what happens when $d(\lambda) = 0$, since the results for this case will be automatically included in the more general results obtained later in this chapter.

3·3. Approximation by kernels of finite rank. In order to extend the results of § 3·2 to more general kernels, we require a theorem on the approximation of $\mathfrak{L}^2$ kernels by kernels of finite rank. We give an outline proof of this theorem here; a second proof, more closely connected with other aspects of the theory of integral equations, will be given later (Theorem 8·3·3).

THEOREM 3·3·1. *If $K(s,t)$ is an $\mathfrak{L}^2$ kernel, and ϵ is an arbitrary positive number, there is an $\mathfrak{L}^2$ kernel $K_0(s,t)$ of finite rank such that $\| K - K_0 \| < \epsilon$.*

We begin by remarking that there is a continuous kernel $K_1(s,t)$ such that $\| K - K_1 \| < \frac{1}{2}\epsilon$; for this result, see, for example, Hobson (1927), p. 638. Next, by the two-dimensional form of Weierstrass's theorem on polynomial approximation (Courant & Hilbert (1953), p. 68), there is a polynomial $P(s,t)$ in the pair of variables (s,t) such that

$$| K_1(s,t) - P(s,t) | < \frac{\epsilon}{2(b-a)} \quad (a \leqslant s \leqslant b,\ a \leqslant t \leqslant b). \tag{1}$$

The polynomial $P(s,t)$ is a kernel of finite rank, and we take $K_0 = P$. We then have

$$\| K_1 - K_0 \|^2 = \iint | K_1(s,t) - P(s,t) |^2 \, ds\, dt \leqslant (b-a)^2 \frac{\epsilon^2}{4(b-a)^2} = \frac{\epsilon^2}{4},$$

whence $\| K_1 - K_0 \| \leqslant \frac{1}{2}\epsilon$, and, by Minkowski's inequality,

$$\| K - K_0 \| \leqslant \| K - K_1 \| + \| K_1 - K_2 \| < \tfrac{1}{2}\epsilon + \tfrac{1}{2}\epsilon = \epsilon,$$

the required result.

The corresponding result for continuous kernels is contained in the inequality (1).

3·4. Reduction of a general equation of the second kind to a finite system of linear equations. Let $K(s, t)$ be an $\mathfrak{L}^2$ kernel, and let ω be a positive number, which may be as large as we please. By Theorem 3·3·1, $K(s,t)$ can be expressed in the form

$$K(s,t) = P(s,t) + Q(s,t), \tag{1}$$

where

$$P(s,t) = \sum_{\nu=1}^{n} a_\nu(s) \overline{b_\nu(t)}$$

is an $\mathfrak{L}^2$ kernel of finite rank, and $\| Q \| < 1/\omega$. We shall suppose throughout that the rank of P is exactly n. Thus K is dissected into two parts, a kernel of finite rank and one of small norm; an integral equation with either as kernel can be dealt with by known methods (cf. § 3·2 for P and § 2·5 for Q), and our next problem is to combine these methods in such a way as to deal with their sum. We shall call an expression of K in the form (1), where P and Q satisfy the conditions stated above, an *ω-dissection* of K.

By Theorem 2·5·1, the kernel $Q(s,t)$ has a resolvent $G_\lambda(s,t)$ for all λ such that $| \lambda | . \| Q \| < 1$, and therefore for $| \lambda | < \omega$; G_λ is given by the relatively uniformly absolutely convergent series

$$G_\lambda = Q + \lambda Q^2 + \lambda^2 Q^3 + \dots;$$

it is a single-valued analytic function of λ for $| \lambda | < \omega$, and it satisfies the resolvent equation

$$G_\lambda - Q = \lambda G_\lambda Q = \lambda Q G_\lambda.$$

To simplify the formulae, we shall write G for G_λ in the rest of the present section and in § 3·5, but its dependence on λ should be borne in mind.

We now show how the general equation of the second kind can be reduced to a finite system of linear equations.

THEOREM 3·4·1. *Let*

$$K(s,t)=P(s,t)+Q(s,t)$$

be an ω-dissection of the $\mathfrak{L}^2$ kernel K, where

$$P=\sum_{\nu=1}^{n} a_\nu \otimes b_\nu$$

is of rank n, and $\| Q \| < 1/\omega$. Let G be the resolvent of Q for a value of λ such that $|\lambda| < \omega$. Write, for $\mathfrak{L}^2$ functions x and y,

$$z=y+\lambda Gy, \quad (x,b_\nu)=x_\nu, \quad (z,b_\nu)=z_\nu,$$

$$(a_\nu+\lambda Ga_\nu, b_\mu)=f_{\mu\nu}.$$

If x satisfies the integral equation

$$x=y+\lambda Kx, \tag{2}$$

then

$$x_\mu=z_\mu+\lambda\sum_{\nu=1}^{n} f_{\mu\nu}x_\nu \quad (1\leqslant\mu\leqslant n). \tag{3}$$

Conversely, if (x_μ) is a solution of (3), *and x is defined by*

$$x=y+\lambda Gy+\lambda\sum_{\nu=1}^{n} x_\nu(a_\nu+\lambda Ga_\nu), \tag{4}$$

then x is an $\mathfrak{L}^2$ solution of (2), *and $x_\nu=(x,b_\nu)$.*

If x satisfies (2), we have

$$x=y+\lambda Kx=(y+\lambda Px)+\lambda Qx.$$

Hence, since Q has G as its resolvent,

$$\begin{aligned}x&=y+\lambda Px+\lambda G(y+\lambda Px)\\&=y+\lambda Gy+\lambda(Px+\lambda GPx)\\&=y+\lambda Gy+\lambda\sum_{\nu=1}^{n}(x,b_\nu)(a_\nu+\lambda Ga_\nu)\\&=z+\lambda Fx,\end{aligned}$$

where

$$F=\sum_{\nu=1}^{n}(a_\nu+\lambda Ga_\nu)\otimes b_\nu=P+\lambda GP \tag{5}$$

is a kernel of finite rank. Conversely, if

$$x = z + \lambda F x = y + \lambda P x + \lambda G(y + \lambda P x),$$

we have, since G is the resolvent of Q,

$$x = y + \lambda P x + \lambda Q x = y + \lambda K x.$$

Thus (2) is equivalent to the equation

$$x = z + \lambda F x, \tag{6}$$

where F is defined by (5). The theorem now follows at once by applying Theorem 3·2·1 to (6).

The numbers z_μ and $f_{\mu\nu}$ are single-valued analytic functions of λ for $|\lambda| < \omega$.

We now turn to the adjoint kernel K^*, and treat it in the same way, obtaining the following result.

THEOREM 3·4·2. *Let*

$$K(s,t) = P(s,t) + Q(s,t)$$

be an ω-dissection of the $\mathfrak{L}^2$ kernel K, where

$$P = \sum_{\nu=1}^{n} a_\nu \otimes b_\nu$$

is of rank n, and $\| Q \| < 1/\omega$. Let G be the resolvent of Q for a value of λ such that $|\lambda| < \omega$. Write, for $\mathfrak{L}^2$ functions u and v,

$$w = v + \bar{\lambda} G^* v, \quad (u, a_\mu) = u_\mu, \quad (w, a_\mu) = w_\mu,$$

$$(a_\nu + \lambda G a_\nu, b_\mu) = f_{\mu\nu}.$$

If u satisfies the integral equation

$$u = v + \bar{\lambda} K^* u, \tag{7}$$

then
$$u_\mu = w_\mu + \bar{\lambda} \sum_{\nu=1}^{n} \overline{f_{\nu\mu}} u_\nu \quad (1 \leqslant \mu \leqslant n). \tag{8}$$

Conversely, if (u_μ) is a solution of (8), *and u is defined by*

$$u = v + \bar{\lambda} G^* v + \bar{\lambda} \sum_{\nu=1}^{n} u_\nu (b_\nu + \bar{\lambda} G^* b_\nu), \tag{9}$$

then u is an $\mathfrak{L}^2$ solution of (7), *and $u_\nu = (u, a_\nu)$.*

We remark that

$$K^* = P^* + Q^* = \sum_{\nu=1}^{n} b_\nu \otimes a_\nu + Q^*$$

is an ω-dissection of K^*, and that G^* is the resolvent of Q^* for the regular value $\bar{\lambda}$. Also

$$\begin{aligned}(b_\nu + \bar{\lambda} G^* b_\nu, a_\mu) &= (b_\nu, a_\mu) + \bar{\lambda}(G^* b_\nu, a_\mu)\\ &= (b_\nu, a_\mu) + (b_\nu, \lambda G a_\mu)\\ &= (b_\nu, a_\mu + \lambda G a_\mu)\\ &= \overline{f_{\nu\mu}}.\end{aligned}$$

Taking these facts into account, we can carry through the proof exactly as for Theorem 3·4·1.

3·5. Meromorphic character of the resolvent. In §3·4 we have shown that the integral equation

$$x = y + \lambda K x \tag{1}$$

and its adjoint equation

$$u = v + \bar{\lambda} K^* u \tag{2}$$

are equivalent respectively to the finite systems of linear equations §3·4 (3) and §3·4 (8), which we may write in matrix form:

$$\mathbf{x} = \mathbf{z} + \lambda \mathbf{F} \mathbf{x}, \tag{3}$$

$$\mathbf{u} = \mathbf{w} + \bar{\lambda} \mathbf{F}^* \mathbf{u}, \tag{4}$$

$\mathbf{F}^*$ being the Hermitian adjoint matrix of $\mathbf{F}$.

We write (3) and (4) as

$$(\mathbf{I} - \lambda \mathbf{F})\, \mathbf{x} = \mathbf{z}, \tag{5}$$

$$(\mathbf{I} - \bar{\lambda} \mathbf{F}^*)\, \mathbf{u} = \mathbf{w}. \tag{6}$$

The determinant $\qquad d(\lambda) = \det(\mathbf{I} - \lambda \mathbf{F})$

is a single-valued analytic function of λ for $|\lambda| < \omega$ (we recall that $\mathbf{F}$ depends on λ); since $d(0) = 1$, it cannot vanish identically, so that it can have only a finite number of zeros in $|\lambda| \leqslant \omega_0$ for any $\omega_0 < \omega$. Leaving these zeros aside for the moment, let us suppose that $|\lambda| < \omega$ and

$$d(\lambda) = \det(\mathbf{I} - \lambda \mathbf{F}) \neq 0,$$

and write $\qquad \mathbf{A} = \operatorname{adj}(\mathbf{I} - \lambda \mathbf{F}) = [\alpha_{\mu\nu}].$

Clearly the elements $\alpha_{\mu\nu}$ of the matrix $\mathbf{A}$ are single-valued analytic functions of λ for $|\lambda|<\omega$. Since $d(\lambda)\neq 0$, (5) has the unique solution

$$\mathbf{x}=\frac{1}{d(\lambda)}\mathbf{Az}, \tag{7}$$

so that, by Theorem 3·4·1, (1) has a unique $\mathfrak{L}^2$ solution x. We proceed to calculate this solution in order to find the form of the resolvent.

By (7), we have

$$x_\mu=\frac{1}{d(\lambda)}\sum_{\nu=1}^{n}\alpha_{\mu\nu}z_\nu=\frac{1}{d(\lambda)}\sum_{\nu=1}^{n}\alpha_{\mu\nu}(y+\lambda Gy, b_\nu),$$

whence, by § 3·4 (4),

$$\begin{aligned} x &= y+\lambda Gy+\frac{\lambda}{d(\lambda)}\sum_{\mu,\nu}\alpha_{\mu\nu}(y+\lambda Gy, b_\nu)(a_\mu+\lambda Ga_\mu) \\ &= y+\lambda Gy+\frac{\lambda}{d(\lambda)}\sum_{\mu,\nu}\alpha_{\mu\nu}(y, b_\nu+\bar{\lambda}G^*b_\nu)(a_\mu+\lambda Ga_\mu) \\ &= y+\lambda H_\lambda y, \end{aligned}$$

where $$H_\lambda=G+\frac{1}{d(\lambda)}\sum_{\mu,\nu}\alpha_{\mu\nu}(a_\mu+\lambda Ga_\mu)\otimes(b_\nu+\bar{\lambda}G^*b_\nu),$$

which is clearly an $\mathfrak{L}^2$ kernel.

These results suggest the following theorem.

THEOREM 3·5·1. *Let $K(s,t)$ be an $\mathfrak{L}^2$ kernel, with the ω-dissection*

$$K=Q+\sum_{\nu=1}^{n}a_\nu\otimes b_\nu,$$

where $\|Q\|<1/\omega$; let G be the resolvent of Q for a value of λ such that $|\lambda|<\omega$, and write $\mathbf{F}=[f_{\mu\nu}]$, where

$$f_{\mu\nu}=(a_\nu+\lambda Ga_\nu, b_\mu) \quad (1\leqslant\mu\leqslant n,\ 1\leqslant\nu\leqslant n);$$

put $$d(\lambda)=\det(\mathbf{I}-\lambda\mathbf{F}),\quad \mathbf{A}=[\alpha_{\mu\nu}]=\operatorname{adj}(\mathbf{I}-\lambda\mathbf{F}).$$

If $|\lambda|<\omega$ and $d(\lambda)\neq 0$, then K has the resolvent H_λ given by

$$H_\lambda=G+\frac{1}{d(\lambda)}\sum_{\mu,\nu=1}^{n}\alpha_{\mu\nu}(a_\mu+\lambda Ga_\mu)\otimes(b_\nu+\bar{\lambda}G^*b_\nu).$$

We verify directly that H_λ satisfies the resolvent equation, observing first that, since

$$(\mathbf{I}-\lambda\mathbf{F})\,\mathbf{A}=(\mathbf{I}-\lambda\mathbf{F})\,\mathrm{adj}\,(\mathbf{I}-\lambda\mathbf{F})=d(\lambda)\,\mathbf{I},$$

we have
$$\lambda\mathbf{F}\mathbf{A}=\mathbf{A}-d(\lambda)\,\mathbf{I},$$

i.e.
$$\lambda\sum_{\mu} f_{\rho\mu}\alpha_{\mu\nu}=\alpha_{\rho\nu}-d(\lambda)\,\delta_{\rho\nu}. \tag{8}$$

We have

$$\lambda K H_\lambda=\lambda(Q+\sum_{\rho} a_\rho\otimes b_\rho)\left[G+\frac{1}{d(\lambda)}\sum_{\mu,\nu}\alpha_{\mu\nu}(a_\mu+\lambda G a_\mu)\otimes(b_\nu+\bar{\lambda}G^*b_\nu)\right]$$

$$=\lambda QG+\lambda\sum_{\rho} a_\rho\otimes G^*b_\rho$$
$$+\frac{\lambda}{d(\lambda)}\sum_{\mu,\nu}\alpha_{\mu\nu}(Qa_\mu+\lambda QGa_\mu)\otimes(b_\nu+\bar{\lambda}G^*b_\nu)$$
$$+\frac{\lambda}{d(\lambda)}\sum_{\rho,\mu,\nu}\alpha_{\mu\nu}(a_\mu+\lambda Ga_\mu,b_\rho)\,a_\rho\otimes(b_\nu+\bar{\lambda}G^*b_\nu)$$
$$=G-Q+\lambda\sum_{\rho} a_\rho\otimes G^*b_\rho$$
$$+\frac{\lambda}{d(\lambda)}\sum_{\mu,\nu}\alpha_{\mu\nu}(Ga_\mu)\otimes(b_\nu+\bar{\lambda}G^*b_\nu)$$
$$+\frac{\lambda}{d(\lambda)}\sum_{\rho,\mu,\nu} f_{\rho\mu}\alpha_{\mu\nu}a_\rho\otimes(b_\nu+\bar{\lambda}G^*b_\nu),$$

since $\lambda QG=G-Q$. Applying (8), we obtain

$$\lambda KH_\lambda=G-Q+\lambda\sum_{\rho} a_\rho\otimes G^*b_\rho$$
$$+\frac{\lambda}{d(\lambda)}\sum_{\mu,\nu}\alpha_{\mu\nu}(Ga_\mu)\otimes(b_\nu+\bar{\lambda}G^*b_\nu)$$
$$+\frac{1}{d(\lambda)}\sum_{\rho,\nu}[\alpha_{\rho\nu}-d(\lambda)\,\delta_{\rho\nu}]\,a_\rho\otimes(b_\nu+\bar{\lambda}G^*b_\nu)$$
$$=G+\frac{1}{d(\lambda)}\sum_{\mu,\nu}\alpha_{\mu\nu}(a_\mu+\lambda Ga_\mu)\otimes(b_\nu+\bar{\lambda}G^*b_\nu)$$
$$-Q+\lambda\sum_{\nu} a_\nu\otimes G^*b_\nu-\sum_{\nu} a_\nu\otimes(b_\nu+\bar{\lambda}G^*b_\nu)$$
$$=H_\lambda-Q-\sum_{\nu} a_\nu\otimes b_\nu$$
$$=H_\lambda-K. \tag{9}$$

The equation $$\lambda H_\lambda K = H_\lambda - K \tag{10}$$
is proved similarly; we need only observe (see end of § 3·4) that
$$f_{\nu\rho} = (a_\rho, b_\nu + \bar{\lambda} G^* b_\nu),$$
and that G^* is the resolvent of Q^* for the regular value $\bar{\lambda}$. The result now follows from (9) and (10).

THEOREM 3·5·2. *Let $K(s,t)$ be an $\mathfrak{L}^2$ kernel. Then the set of values of λ for which the resolvent H_λ of K fails to exist is at most enumerable, and has no finite limit point, and the resolvent kernel $H_\lambda(s,t)$ is a meromorphic function of λ.*

It follows at once from Theorem 3·5·1 that $H_\lambda(s,t)$ is a single-valued analytic function of λ in the domain $|\lambda| < \omega$, except for poles at the zeros of $d(\lambda)$; since the number ω used in forming the ω-dissection of K can be chosen as large as we please, the only singularities of $H_\lambda(s,t)$ in the whole plane are poles, and, since there is only a finite number of these in $|\lambda| \leqslant \omega_0$ for any ω_0, they are at most enumerable and have no finite limit point. The resolvent kernel $H_\lambda(s,t)$ is thus a meromorphic function of λ for all (s,t).

3·6. The homogeneous equation. We have seen in § 3·5 that when (in the notation of Theorem 3·5·1)
$$\det(\mathbf{I} - \lambda \mathbf{F}) = d(\lambda) \neq 0,$$
λ is a regular value of the $\mathfrak{L}^2$ kernel K. We now turn our attention to what happens when $d(\lambda) = 0$.

THEOREM 3·6·1. *Let $K(s,t)$ be an $\mathfrak{L}^2$ kernel. Then*

(a) every complex number λ is either a regular value or a characteristic value of K; in other words, either the equation
$$x(s) = y(s) + \lambda \int K(s,t)\, x(t)\, dt \tag{1}$$
has a unique $\mathfrak{L}^2$ solution $x(s)$ for every $\mathfrak{L}^2$ function $y(s)$, or the associated homogeneous equation
$$x(s) = \lambda \int K(s,t)\, x(t)\, dt \tag{2}$$
has an $\mathfrak{L}^2$ solution $x(s)$ that does not vanish identically.

(b) When λ is a characteristic value of K, equation (2) has at most a finite number of linearly independent $\mathfrak{L}^2$ solutions, and the number of functions in any maximal linearly independent set of solutions is equal to the corresponding number for the adjoint equation

$$u(s)=\bar{\lambda}\int\overline{K(t,s)}\,u(t)\,dt. \tag{3}$$

We choose ω so that $|\lambda|<\omega$, and perform an ω-dissection of K. If (1) fails to have a unique $\mathfrak{L}^2$ solution for some $\mathfrak{L}^2$ function $y(s)$, it follows from Theorem 3·4·1 that the system of equations

$$x_\mu=z_\mu+\lambda\sum_{\nu=1}^{n}f_{\mu\nu}x_\nu \quad (1\leqslant\mu\leqslant n)$$

does not have a unique solution; this happens if and only if

$$d(\lambda)=\det(\mathbf{I}-\lambda\mathbf{F})=0.$$

Let r be the rank of $\mathbf{I}-\lambda\mathbf{F}$; since $r<n$, $p=n-r>0$, and the vector equation

$$(\mathbf{I}-\lambda\mathbf{F})\,\mathbf{x}=\mathbf{0} \tag{4}$$

has p linearly independent solutions $\mathbf{x}^{(1)},\mathbf{x}^{(2)},\ldots,\mathbf{x}^{(p)}$. The general solution of (4) is therefore of the form

$$\mathbf{x}=\sum_{\sigma=1}^{p}\alpha_\sigma\mathbf{x}^{(\sigma)},$$

the coefficients α_σ being arbitrary complex numbers. The correspondence between $\mathfrak{L}^2$ solutions x of (2) and solutions $\mathbf{x}$ of (4) is expressed by the equations

$$x=\lambda\sum_{\nu=1}^{n}x_\nu(a_\nu+\lambda Ga_\nu),\quad x_\nu=(x,b_\nu)\quad(1\leqslant\nu\leqslant n);$$

since this relation is linear and (1, 1), the general $\mathfrak{L}^2$ solution of (2) is of the form

$$x=\sum_{\sigma=1}^{p}\alpha_\sigma x^{(\sigma)},$$

where $x^{(\sigma)}$ is the solution of (2) corresponding to $\mathbf{x}^{(\sigma)}$, and the functions $x^{(\sigma)}$ are linearly independent. Thus λ is a characteristic value of K, and its characteristic subspace is finite-dimensional, the functions $x^{(1)},\ldots,x^{(p)}$ forming a base for it. We call the integer p

the *rank* of the characteristic value λ, and a base of the characteristic subspace of λ will be called a *full system* of characteristic functions belonging to λ.

Since $\mathbf{I}-\bar{\lambda}\mathbf{F}^*=(\mathbf{I}-\lambda\mathbf{F})^*$ has the same rank as $\mathbf{I}-\lambda\mathbf{F}$, the vector equation

$$(\mathbf{I}-\bar{\lambda}\mathbf{F}^*)\mathbf{u}=\mathbf{0}$$

also has p linearly independent solutions, which give rise to a full system of p characteristic functions of K^* belonging to the characteristic value $\bar{\lambda}$. Thus λ and $\bar{\lambda}$ have the same rank, and the proof of the theorem is complete.

3·7. The non-homogeneous equation when the parameter is a characteristic value. We now consider the non-homogeneous equation

$$x=y+\lambda Kx$$

in the case when λ is a characteristic value of the $\mathfrak{L}^2$ kernel K. The equation may or may not have an $\mathfrak{L}^2$ solution x; when it does, we have the following result, whose proof is now quite elementary.

THEOREM 3·7·1. *If the characteristic value λ of the $\mathfrak{L}^2$ kernel $K(s,t)$ has rank p, and the equation*

$$x(s)=y(s)+\lambda\int K(s,t)\,x(t)\,dt \tag{1}$$

has an $\mathfrak{L}^2$ solution $x_0(s)$ for a given $\mathfrak{L}^2$ function $y(s)$, then the general $\mathfrak{L}^2$ solution of (1) *is given by*

$$x(s)=x_0(s)+\sum_{\sigma=1}^{p}\alpha_\sigma x^{(\sigma)}(s), \tag{2}$$

where $\{x^{(\sigma)}(s)\}$ is a full system of characteristic functions of $K(s,t)$ for the characteristic value λ.

We have

$$x_0=y+\lambda Kx_0,$$

and, if x is given by (2),

$$\begin{aligned}y+\lambda Kx&=y+\lambda Kx_0+\sum_{\sigma=1}^{p}\alpha_\sigma\lambda Kx^{(\sigma)}\\&=x_0+\sum_{\sigma=1}^{p}\alpha_\sigma x^{(\sigma)}\\&=x,\end{aligned}$$

so that x satisfies (1), and it is an $\mathfrak{L}^2$ function. Conversely, if x is an arbitrary $\mathfrak{L}^2$ solution of (1), then

$$x = y + \lambda K x, \quad x_0 = y + \lambda K x_0;$$

subtracting, we have

$$x - x_0 = \lambda K(x - x_0),$$

so that $x - x_0$ satisfies the homogeneous equation associated with (1), and is therefore a linear combination of the $x^{(\sigma)}$; hence x is of the form (2).

THEOREM 3·7·2. *If $K(s,t)$ is an $\mathfrak{L}^2$ kernel, and $y(s)$ is an $\mathfrak{L}^2$ function, a necessary and sufficient condition for the existence of an $\mathfrak{L}^2$ solution $x(s)$ of the integral equation*

$$x(s) = y(s) + \lambda \int K(s,t)\, x(t)\, dt \tag{3}$$

is that y be orthogonal to every $\mathfrak{L}^2$ solution u of the adjoint homogeneous equation

$$u(s) = \bar{\lambda} \int \overline{K(t,s)}\, u(t)\, dt. \tag{4}$$

If λ is a regular value of K, there is nothing to prove. If λ is a characteristic value, we perform an ω-dissection of K for some $\omega > |\lambda|$, and carry out the construction of § 3·4. Equation (3) is then equivalent to

$$(\mathbf{I} - \lambda \mathbf{F})\, \mathbf{x} = \mathbf{z}, \tag{5}$$

where $\mathbf{I} - \lambda \mathbf{F}$ is of rank $r < n$. If $\mathbf{u}$ satisfies

$$(\mathbf{I} - \bar{\lambda} \mathbf{F}^*)\, \mathbf{u} = \mathbf{0}, \tag{6}$$

we have, by transposing and taking the complex conjugate,

$$\mathbf{u}^*(\mathbf{I} - \lambda \mathbf{F}) = \mathbf{0}^*, \tag{7}$$

where $\mathbf{0}^*$ is a row matrix consisting of 0's. From (5) and (7) we obtain

$$\mathbf{u}^*\mathbf{z} = \mathbf{u}^*(\mathbf{I} - \lambda \mathbf{F})\, \mathbf{x} = \mathbf{0}^*\mathbf{x} = 0;$$

thus a necessary condition for (5) to have a solution is that $\mathbf{u}^*\mathbf{z} = 0$ for every $\mathbf{u}$ satisfying (6). Since the subspace of solutions $\mathbf{u}$ of (6) is of dimension $p = n - r$, the vectors $\mathbf{z}$ satisfying this condition form a subspace M of dimension r. On the other hand, vectors of the form $(\mathbf{I}-\lambda\mathbf{F})\mathbf{x}$, where $\mathbf{x}$ is arbitrary, form a subspace N of dimension r contained in M, and therefore identical with M; in other words, the condition is sufficient for the existence of a solution of (5).

We have seen that there is a linear (1, 1) correspondence between $\mathfrak{L}^2$ solutions u of (4) and solutions $\mathbf{u}=(u_\nu)$ of (6), given by

$$u_\nu=(u,a_\nu)\quad(1\leqslant\nu\leqslant n),$$

$$u=\bar{\lambda}\sum_{\nu=1}^{n}u_\nu(b_\nu+\bar{\lambda}G^*b_\nu);$$

we then have
$$(y,u)=\lambda\sum_{\nu=1}^{n}\bar{u}_\nu(y,b_\nu+\bar{\lambda}G^*b_\nu)$$
$$=\lambda\sum_{\nu=1}^{n}\bar{u}_\nu(y+\lambda Gy,b_\nu)$$
$$=\lambda\sum_{\nu=1}^{n}\bar{u}_\nu z_\nu$$
$$=\lambda\mathbf{u}^*\mathbf{z},$$

since $(y+\lambda Gy,b_\nu)=z_\nu$. Thus $\mathbf{u}^*\mathbf{z}=0$ for every solution $\mathbf{u}$ of (6) if and only if $(y,u)=0$ for every $\mathfrak{L}^2$ solution u of (4), and the theorem is proved.

The necessity of the condition can also be proved directly: if x and u are $\mathfrak{L}^2$ solutions of (3) and (4) respectively, we have

$$(y,u)=(x-\lambda Kx,u)=(x,u-\bar{\lambda}K^*u)=0.$$

3·8. Concluding remarks. We may summarize our main results as follows:

If K is an $\mathfrak{L}^2$ kernel, then

(a) every complex number λ is either a regular value or a characteristic value of K; the set of characteristic values is at most enumerable and has no finite limit point, and the resolvent kernel H_λ of K is a meromorphic function of λ;

(b) if λ is a regular value of K, then $\bar{\lambda}$ is a regular value of K^*, and the equations

$$x=y+\lambda Kx,\quad u=v+\bar{\lambda}K^*u,$$

have unique $\mathfrak{L}^2$ solutions x and u for any $\mathfrak{L}^2$ functions y and v, given by
$$x=y+\lambda H_\lambda y,\quad u=v+\bar{\lambda}(H_\lambda)^*v;$$

(c) *if λ is a characteristic value of K, and has rank p, then $\bar{\lambda}$ is a characteristic value of K^* of rank p;*

(d) *if λ is a characteristic value of K and y is an $\mathfrak{L}^2$ function, the equation*

$$x = y + \lambda K x \tag{1}$$

has an $\mathfrak{L}^2$ solution x if and only if y is orthogonal to every $\mathfrak{L}^2$ solution u of the adjoint homogeneous equation $u = \bar{\lambda} K^* u$; and the general $\mathfrak{L}^2$ solution of (1) is obtained by adding to a particular $\mathfrak{L}^2$ solution the general $\mathfrak{L}^2$ solution of the homogeneous equation $x = \lambda K x$.

The statements (*b*)–(*d*), taken together, are sometimes known as the 'determinant-free' theorems, since they do not involve the Fredholm determinant and its minors, which we shall discuss in Chapters V and VI; the statement that every complex number is either a regular value or a characteristic value expresses the 'Fredholm alternative'.

We observe that the determinant-free theorems are closely analogous to the fundamental results about finite systems of linear equations in a finite number of unknowns; this should not now be surprising, since our leading idea has been to reduce the integral equation to an equivalent system of this kind. It is essential to the applicability of the process that the integral equations considered are of the second kind; no such reduction exists for equations of the first kind, nor is there an analogous group of theorems.

The adaptation of the results of the present chapter to the case of continuous kernels and functions may be left to the reader; no essential changes are required.

The method of the present chapter can be used in certain cases for the numerical solution of integral equations; for details, see, for example, Bückner (1952).

CHAPTER IV

ORTHONORMAL SYSTEMS OF FUNCTIONS

4·1. Definitions and elementary properties. Before we take the theory of integral equations any farther, it is convenient to insert at this point a brief account of orthonormal systems of functions, which will be used a good deal in the rest of the book.

We recall (§ 1·4) that the $\mathfrak{L}^2$ functions x and y are said to be orthogonal to one another if $(x, y) = 0$. A set $\mathfrak{S}$ of $\mathfrak{L}^2$ functions is called an *orthonormal*† *system* if (i) any two different functions x and y of $\mathfrak{S}$ are orthogonal, and (ii) $\| x \| = 1$ for every x of $\mathfrak{S}$.

As examples of orthonormal systems we may mention:

(*a*) the trigonometrical system $e^{2\pi i n t}$ $(n = 0, \pm 1, \pm 2, \ldots)$ in the interval $0 \leqslant t \leqslant 1$;

(*b*) the Legendre system $\sqrt{(n + \frac{1}{2})}\, P_n(t)$ $(n = 0, 1, 2, \ldots)$ in the interval $-1 \leqslant t \leqslant 1$, where $P_n(t)$ denotes the Legendre polynomial of degree n.

An orthonormal system is said to be *complete* if it is not a proper subset of a more comprehensive orthonormal system. The system $\mathfrak{S}$ is complete if and only if every $\mathfrak{L}^2$ function orthogonal to every member of $\mathfrak{S}$ is a null function. For, if this condition is satisfied, it is clear that $\mathfrak{S}$ cannot have a proper orthonormal extension; and if there is an $\mathfrak{L}^2$ function y orthogonal to every member of $\mathfrak{S}$ and such that $\| y \| \neq 0$, the adjunction of $y / \| y \|$ to $\mathfrak{S}$ gives a proper orthonormal extension. We remark at this point that the notion of the equivalence of two functions, i.e. their equality almost everywhere, will play a larger part from now on than it has done hitherto.

The two orthonormal systems given as examples above are both known to be complete. We shall not require the theory of complete orthonormal systems in this book, except in the proof of one rather special result (Theorem 8·6·7) and in some incidental remarks. We quote the following results‡ without proof: every orthonormal system is finite or enumerable; a

† The word 'orthonormal' is an abbreviation for 'orthogonal and normal'.

‡ See, for example, Kaczmarz & Steinhaus (1935).

complete system is always enumerable; an incomplete system can always be completed by adjoining a finite or enumerable set of $\mathfrak{L}^2$ functions.

From now on we shall suppose that every orthonormal system considered is finite or enumerable, and we shall usually write it as a sequence $(e_1, e_2, \ldots)$.

Any finite subset $e_1, \ldots, e_n$ of an orthonormal system is linearly independent with respect to equivalence; for, if

$$\alpha_1 e_1 + \alpha_2 e_2 + \ldots + \alpha_n e_n =^\circ 0, \tag{1}$$

we can form the inner product of (1) with any e_m, where $1 \leqslant m \leqslant n$, obtaining

$$\sum_{\nu=1}^{n} \alpha_\nu (e_\nu, e_m) = \alpha_m (e_m, e_m) = \alpha_m = 0.$$

We remark that a system of functions that is linearly independent with respect to equivalence is *a fortiori* linearly independent in the ordinary sense; the converse is false, as can be seen by considering the function 0 and a null function that does not vanish identically.

4·2. Mean square approximation and Bessel's inequality. Let $(e_1, e_2, \ldots, e_n)$ be a finite orthonormal system and x an arbitrary $\mathfrak{L}^2$ function. We consider the problem of finding the best mean square approximation to x by a linear combination of $e_1, e_2, \ldots, e_n$, i.e. of choosing the coefficients $\alpha_1, \alpha_2, \ldots, \alpha_n$ in such a way as to make the expression

$$\left\| x - \sum_{\nu=1}^{n} \alpha_\nu e_\nu \right\| \tag{1}$$

a minimum.

THEOREM 4·2·1. *For a given $\mathfrak{L}^2$ function x and a given orthonormal system $(e_1, e_2, \ldots, e_n)$ the expression* (1) *attains its minimum value when*

$$\alpha_\nu = \xi_\nu = (x, e_\nu) \quad (1 \leqslant \nu \leqslant n).$$

We then have

$$\sum_{\nu=1}^{n} |\xi_\nu|^2 \leqslant \|x\|^2. \tag{2}$$

Writing $\xi_\nu = (x, e_\nu)$ $(1 \leqslant \nu \leqslant n)$, we have

$$\left\| x - \sum_{\nu=1}^{n} \alpha_\nu e_\nu \right\|^2 = \left(x - \sum_{\mu=1}^{n} \alpha_\mu e_\mu, \, x - \sum_{\nu=1}^{n} \alpha_\nu e_\nu \right)$$

$$= (x, x) - \sum_{\mu} \alpha_\mu (e_\mu, x) - \sum_{\nu} \bar{\alpha}_\nu (x, e_\nu) + \sum_{\mu, \nu} \alpha_\mu \bar{\alpha}_\nu (e_\mu, e_\nu)$$

$$= \| x \|^2 - \sum_{\nu} \alpha_\nu \bar{\xi}_\nu - \sum_{\nu} \bar{\alpha}_\nu \xi_\nu + \sum_{\nu} \alpha_\nu \bar{\alpha}_\nu$$

$$= \| x \|^2 - \sum_{\nu} \xi_\nu \bar{\xi}_\nu + \sum_{\nu} (\xi_\nu - \alpha_\nu)(\bar{\xi}_\nu - \bar{\alpha}_\nu)$$

$$= \| x \|^2 - \sum_{\nu} | \xi_\nu |^2 + \sum_{\nu} | \xi_\nu - \alpha_\nu |^2. \tag{3}$$

It is evident from (3) that the right-hand side attains its minimum value when (and only when) $\alpha_\nu = \xi_\nu$ $(1 \leqslant \nu \leqslant n)$; we then have

$$\left\| x - \sum_{\nu=1}^{n} \xi_\nu x_\nu \right\|^2 = \| x \|^2 - \sum_{\nu=1}^{n} | \xi_\nu |^2. \tag{4}$$

Since the left-hand side of (4) cannot be negative, the inequality (2) follows at once; it is usually known as *Bessel's inequality.*

We remark that the number ξ_ν depends only on x and e_ν, and not on the other members of the orthonormal system.

COROLLARY 1. *If* (e_n) *is an enumerable orthonormal system,* x *is an* $\mathfrak{L}^2$ *function, and*

$$\xi_n = (x, e_n) \quad (n = 1, 2, \ldots),$$

then the series $\sum_{n=1}^{\infty} | \xi_n |^2$ *is convergent, and*

$$\sum_{n=1}^{\infty} | \xi_n |^2 \leqslant \| x \|^2. \tag{5}$$

COROLLARY 2. *If* (e_n) *is an orthonormal system,* x *and* y *are* $\mathfrak{L}^2$ *functions, and*

$$\xi_n = (x, e_n), \quad \eta_n = (y, e_n) \quad (n = 1, 2, \ldots),$$

then

$$\left| \sum_{n} \xi_n \bar{\eta}_n \right| \leqslant \| x \| . \| y \|, \tag{6}$$

the series on the left-hand side of (6) *being absolutely convergent if the system* (e_n) *is infinite.*

For, by Cauchy's inequality (§ 1·4),

$$\left|\sum_{\nu=1}^{n}\xi_\nu\bar{\eta}_\nu\right| \leqslant \sum_{\nu=1}^{n}|\xi_\nu\bar{\eta}_\nu| \leqslant \left(\sum_{\nu=1}^{n}|\xi_\nu|^2\right)^{\frac{1}{2}}\left(\sum_{\nu=1}^{n}|\eta_\nu|^2\right)^{\frac{1}{2}}.$$

The numbers $\xi_n=(x, e_n)$ are called the *Fourier coefficients* of x with respect to the orthonormal system (e_n); we sometimes write $x \sim (\xi_n)$ or

$$x \sim \sum_n \xi_n e_n.$$

For instance, the Fourier coefficients of an $\mathfrak{L}^2$ function $x(t)$ with respect to the orthonormal system $e^{2\pi int}$ $(n=0, \pm 1, \pm 2, \ldots)$ are

$$\xi_n=\int_0^1 x(t)\, e^{-2\pi int}\, dt,$$

which are just the ordinary trigonometrical Fourier coefficients.

4·3. Mean square convergence and the Riesz–Fischer theorem. In §4·2 we have defined the Fourier coefficients of an $\mathfrak{L}^2$ function with respect to an orthonormal system; we now ask under what conditions a given sequence of complex numbers (ξ_n) is the sequence of Fourier coefficients of some $\mathfrak{L}^2$ function with respect to a given orthonormal system (e_n). If the given sequences are finite, say $(\xi_1, \ldots, \xi_p)$ and $(e_1, \ldots, e_p)$ respectively, we can give an immediate answer; such a function always exists, e.g.

$$x=\xi_1 e_1+\ldots+\xi_p e_p.$$

For infinite sequences the answer, as we shall see in a moment, is simple but non-trivial.

Since

$$\| x-y \| = \| y-x \|,$$

and $$\| x-z \| = \| (x-y)+(y-z) \| \leqslant \| x-y \| + \| y-z \|,$$

the expression $\| x-y \|$ has two of the properties usually required for a distance function or *metric*; it lacks the third, since $\| x-y \|$ may vanish without x and y being identical. However, $\| x-y \|$ can be usefully regarded as defining a kind of distance, which we call the *mean square distance*, between the $\mathfrak{L}^2$ functions x and y.

We are thus led to introduce the notion of convergence with respect to this distance; a sequence (x_n) of $\mathfrak{L}^2$ functions is said to be *convergent in mean square* to the $\mathfrak{L}^2$ function x if

$$\lim_{n\to\infty} \| x_n - x \| = 0,$$

and x is called the *mean square limit* of (x_n). This type of convergence is also called *strong convergence* (with *index* 2) or *convergence in* $\mathfrak{L}^2$, and we write $x_n \to x(\mathfrak{L}^2)$.

If $x_n \to x(\mathfrak{L}^2)$ and $x = {}^\circ y$, then $x_n \to y(\mathfrak{L}^2)$; conversely, if $x_n \to x(\mathfrak{L}^2)$ and $x_n \to y(\mathfrak{L}^2)$, then

$$\| x - y \| \leqslant \| x - x_n \| + \| x_n - y \| \to 0 \quad (n \to \infty),$$

whence $\| x - y \| = 0$, $x = {}^\circ y$. Thus mean square limits are unique, but only in the sense of equivalence.

An infinite series

$$\sum_{n=1}^{\infty} x_n = x_1 + x_2 + \dots$$

of $\mathfrak{L}^2$ functions is said to be convergent in mean square to the sum s if

$$s_n = x_1 + \dots + x_n \to s\,(\mathfrak{L}^2).$$

We write also

$$\sum_{n=1}^{\infty} x_n = {}^\circ s\,(\mathfrak{L}^2).$$

We now show that the expressions λx, $x + y$, (x, y) and $\| x \|$ are continuous with respect to convergence in mean square.

(i) If $\lambda_n \to \lambda$ and $x_n \to x(\mathfrak{L}^2)$, then

$$\begin{aligned} \| \lambda_n x_n - \lambda x \| &= \| (\lambda_n - \lambda)(x_n - x) + \lambda(x_n - x) + (\lambda_n - \lambda) x \| \\ &\leqslant | \lambda_n - \lambda | . \| x_n - x \| + | \lambda | . \| x_n - x \| \\ &\qquad + | \lambda_n - \lambda | . \| x \| \to 0, \end{aligned}$$

i.e. $\lambda_n x_n \to \lambda x(\mathfrak{L}^2)$.

(ii) If $x_n \to x(\mathfrak{L}^2)$ and $y_n \to y(\mathfrak{L}^2)$, then

$$\| (x_n + y_n) - (x + y) \| \leqslant \| x_n - x \| + \| y_n - y \| \to 0,$$

i.e. $x_n + y_n \to x + y(\mathfrak{L}^2)$.

(iii) If $x_n \to x(\mathfrak{L}^2)$ and $y_n \to y(\mathfrak{L}^2)$, we see by an argument like that in (i) that $(x_n, y_n) \to (x, y)$. In particular, if $x_n \to x(\mathfrak{L}^2)$, then

$$\| x_n \|^2 \to \| x \|^2, \; \| x_n \| \to \| x \|.$$

It can be shown by examples that convergence in mean square neither implies nor is implied by convergence almost everywhere, even if we keep within the class of $\mathfrak{L}^2$ functions. However, if the sequence (x_n) is relatively uniformly convergent to x, then, for some non-negative $\mathfrak{L}^2$ function $p(t)$,

$$|x_n(t)-x(t)| \leqslant \epsilon p(t) \quad (n > n_0),$$

so that $$\int |x_n(t)-x(t)|^2 dt \leqslant \epsilon^2 \int [p(t)]^2 dt,$$

i.e., $$\|x_n - x\| \leqslant \epsilon \|p\|,$$

whence $x_n \to x(\mathfrak{L}^2)$.

A sequence (x_n) of $\mathfrak{L}^2$ functions is called a *Cauchy sequence* if

$$\lim_{m,n\to\infty} \|x_m - x_n\| = 0.$$

Every sequence that is convergent in mean square is a Cauchy sequence; for, if $x_n \to x(\mathfrak{L}^2)$, then

$$\|x_m - x_n\| \leqslant \|x_m - x\| + \|x - x_n\| \to 0 \quad (m, n \to \infty).$$

The converse is also true, as is shown by the following theorem, which is analogous to the 'general principle of convergence' for sequences of real or complex numbers. It may be regarded as expressing the completeness, in the topological sense, of the space of $\mathfrak{L}^2$ functions when $\|x-y\|$ is used as a distance function. The validity of the theorem depends essentially on the fact that we are using the Lebesgue integral; no such result holds, for instance, for functions of Riemann integrable square.

THEOREM 4·3·1. *If (x_n) is a Cauchy sequence of $\mathfrak{L}^2$ functions, there is an $\mathfrak{L}^2$ function x such that $x_n \to x(\mathfrak{L}^2)$.*

For a proof, we refer the reader to Burkill (1951), pp. 67–8. The main idea of the usual proof (Weyl†) is to extract from (x_n) a subsequence that converges almost everywhere to an $\mathfrak{L}^2$ function x, and then to show that x is the $\mathfrak{L}^2$ limit of (x_n). Theorem 4·3·1 is Fischer's form of the Riesz–Fischer theorem.‡

† Weyl (1909).
‡ F. Riesz (1907*a*, *b*), Fischer (1907).

From Theorem 4·3·1 we deduce the following result, which is F. Riesz's form of the Riesz–Fischer theorem.

THEOREM 4·3·2. *Given an enumerable orthonormal system* (e_n) *and a sequence* (ξ_n) *of complex numbers, a necessary and sufficient condition for the infinite series*

$$\sum_{n=1}^{\infty} \xi_n e_n \tag{1}$$

to be convergent in mean square is that

$$\sum_{n=1}^{\infty} |\xi_n|^2 < \infty. \tag{2}$$

Furthermore, if

$$x = {}^{\circ}\sum_{n=1}^{\infty} \xi_n e_n \ (\mathfrak{L}^2),$$

then

$$x \sim (\xi_n)$$

and

$$\|x\|^2 = \sum_{n=1}^{\infty} |\xi_n|^2. \tag{3}$$

If (1) is convergent in mean square, its partial sums form a Cauchy sequence, i.e.

$$\left\| \sum_{\nu=m+1}^{n} \xi_\nu e_\nu \right\| \to 0 \quad (m, n \to \infty).$$

Now

$$\left\| \sum_{\nu=m+1}^{n} \xi_\nu e_\nu \right\|^2 = \left(\sum_{\mu=m+1}^{n} \xi_\mu e_\mu, \sum_{\nu=m+1}^{n} \xi_\nu e_\nu \right)$$

$$= \sum_{\mu, \nu=m+1}^{n} \xi_\mu \bar{\xi}_\nu (e_\mu, e_\nu)$$

$$= \sum_{\nu=m+1}^{n} \xi_\nu \bar{\xi}_\nu = \sum_{\nu=m+1}^{n} |\xi_\nu|^2, \tag{4}$$

since (e_n) is orthonormal. Thus

$$\sum_{\nu=m+1}^{n} |\xi_\nu|^2 \to 0 \quad (m, n \to \infty),$$

whence, by the general principle of convergence for series of real terms, (2) must hold.

Conversely, if (2) holds, the identity (4) shows that the partial sums

$$s_n = \xi_1 e_1 + \ldots + \xi_n e_n$$

form a Cauchy sequence, which is convergent, by Theorem 4·3·1, to some $\mathfrak{L}^2$ function x. We then have

$$(s_n, e_\nu) = \xi_\nu \quad (n \geqslant \nu),$$

whence $$(x, e_\nu) = \lim_{n\to\infty} (s_n, e_\nu) = \lim_{n\to\infty} \xi_\nu = \xi_\nu \quad (\nu = 1, 2, \ldots),$$

i.e. $$x \sim (\xi_n).$$

Finally, $$\| x \|^2 = \lim_{n\to\infty} \| s_n \|^2 = \lim_{n\to\infty} \sum_{\nu=1}^{n} | \xi_\nu |^2 = \sum_{n=1}^{\infty} | \xi_n |^2.$$

This completes the proof.

We note that the necessary and sufficient condition (2) does not depend on the particular orthonormal system (e_n).

COROLLARY. *If*

$$x = {}^\circ\sum_{n=1}^{\infty} \xi_n e_n \quad (\mathfrak{L}^2), \qquad y = {}^\circ\sum_{n=1}^{\infty} \eta_n e_n \quad (\mathfrak{L}^2),$$

then $$(x, y) = \sum_{n=1}^{\infty} \xi_n \overline{\eta}_n.$$

Writing $$s_n = \xi_1 e_1 + \ldots + \xi_n e_n,$$
$$t_n = \eta_1 e_1 + \ldots + \eta_n e_n,$$

we have $$(x, y) = \lim_{n\to\infty} (s_n, t_n) = \lim_{n\to\infty} \sum_{\nu=1}^{n} \xi_\nu \overline{\eta}_\nu = \sum_{n=1}^{\infty} \xi_n \overline{\eta}_n.$$

We remark that if (e_n) is a complete orthonormal system and x is an arbitrary $\mathfrak{L}^2$ function, then

$$x = {}^\circ\sum_{n=1}^{\infty} (x, e_n) e_n \quad (\mathfrak{L}^2) \tag{5}$$

and $$\| x \|^2 = \sum_{n=1}^{\infty} | (x, e_n) |^2. \tag{6}$$

More generally, if x and y are arbitrary $\mathfrak{L}^2$ functions, then

$$(x, y) = \sum_{n=1}^{\infty} (x, e_n)(e_n, y). \tag{7}$$

For, by Theorem 4·3·2, the series on the right-hand side of (5) is convergent in mean square to an $\mathfrak{L}^2$ function y such that $(y, e_n) = (x, e_n)$, i.e. $(y - x, e_n) = 0$ $(n \geq 1)$. Since (e_n) is complete, we have $y - x =^\circ 0$, $y =^\circ x$. Equation (6), which is known as the *Parseval formula*, then follows from Theorem 4·3·2; and equation (7), the *generalized Parseval formula*, from the Corollary to Theorem 4·3·2.

4·4. The orthogonalization process. We shall say that two sequences of $\mathfrak{L}^2$ functions are *linearly equivalent* to one another if every member of each sequence is equivalent (i.e. equal almost everywhere) to a finite linear combination of members of the other sequence. We often want to replace a given sequence of $\mathfrak{L}^2$ functions by a linearly equivalent orthonormal system; the following result enables us to do so.

THEOREM 4·4·1. *Let (x_n) be a finite or infinite sequence of $\mathfrak{L}^2$ functions, not all of which are null. Then there exists an orthonormal system (e_n) linearly equivalent to (x_n).*

We begin by omitting from the sequence (x_n) any function that is equivalent to a linear combination of earlier members of the sequence; thus we omit x_1 if it is a null function, x_2 if it is equivalent to a scalar multiple of x_1, and so on. The remaining functions form a sequence (y_n), which is clearly linearly equivalent to (x_n), and also linearly independent with respect to equivalence.

We now construct the sequence (e_n) inductively. We take

$$e_1 = y_1 / \| y_1 \|,$$

as we can do since $\| y_1 \| \neq 0$. Suppose that we have constructed $(e_1, e_2, \ldots, e_p)$, orthonormal and linearly equivalent to $(y_1, y_2, \ldots, y_p)$, and that we have not exhausted the sequence (y_n), so that y_{p+1} exists. Write

$$u_{p+1} = y_{p+1} - (y_{p+1}, e_1)\, e_1 - \ldots - (y_{p+1}, e_p)\, e_p. \tag{1}$$

Then u_{p+1} is equivalent to (actually equal to) a finite linear combination of $y_1, \ldots, y_p$ and y_{p+1}; also $\| u_{p+1} \| \neq 0$, for if u_{p+1}

were null, y_{p+1} would be equivalent to a linear combination of $y_1, \ldots, y_p$, which is impossible. Also, if $1 \leqslant r \leqslant p$, we have

$$(u_{p+1}, e_r) = (y_{p+1}, e_r) - \sum_{\rho=1}^{p} (y_{p+1}, e_\rho)(e_\rho, e_r)$$

$$= (y_{p+1}, e_r) - (y_{p+1}, e_r) = 0,$$

since $(e_1, \ldots, e_p)$ is orthonormal; thus u_{p+1} is orthogonal to $e_1, e_2, \ldots, e_p$. The same is therefore true of

$$e_{p+1} = u_{p+1}/\| u_{p+1} \|,$$

and $\| e_{p+1} \| = 1$, so that $(e_1, \ldots, e_p, e_{p+1})$ is orthonormal and, by (1) and the inductive hypothesis, linearly equivalent to $(y_1, \ldots, y_p, y_{p+1})$. The induction can therefore be carried through. Since the full sequences (e_n) and (y_n) are linearly equivalent, the same is true for (e_n) and (x_n).

The process that we have just described is known as the Gram–Schmidt† orthogonalization process.

We cannot in general replace equivalence by equality in Theorem 4·4·1, as we can see by taking $\| x_1 \| = \| x_2 \| = 1$, $x_1 = {}^\circ x_2$, $x_1 \neq x_2$; application of the orthogonalization process to these two functions gives the one-element orthonormal system (x_1), and it is clear that there is no orthonormal system of which both x_1 and x_2 are finite linear combinations. We can, however, replace equivalence by equality if the given sequence (x_n) is already linearly independent with respect to equivalence; we can also do so if all the functions x_n are continuous, and the orthonormal system (e_n) will then consist of continuous functions.

4·5. Orthonormal systems of functions of two variables. The whole of the discussion of orthonormal systems in §§ 4·1–4·4 can be carried through for functions of two variables, and we shall consider ourselves at liberty to quote results such as Bessel's inequality or the Riesz–Fischer theorem for such systems. Some new points arise, however, and we give below some special results that we shall need later.

Let (e_n) and (f_n) be two orthonormal systems with the same

† Gram (1883), Schmidt (1907*a*).

number of elements; then $(e_n \otimes f_n)$ is an orthonormal system of functions of two variables. For

$$\begin{aligned}(e_m \otimes f_m, e_n \otimes f_n) &= \iint e_m(s)\overline{f_m(t)}\,\overline{e_n(s)}\,f_n(t)\,ds\,dt \\ &= \int e_m(s)\overline{e_n(s)}\,ds \int \overline{f_m(t)}\,f_n(t)\,dt \\ &= (e_m, e_n)(f_n, f_m) \\ &= \delta_{mn}.\end{aligned}$$

The following result will be used in Chapters VII and VIII.

THEOREM 4·5·1. *Let* (e_n) *and* (f_n) *be infinite orthonormal systems, and form the orthonormal system* $(e_n \otimes f_n)$. *Let* (κ_n) *be a sequence of complex numbers such that*

$$\sum_{n=1}^{\infty} |\kappa_n|^2 < \infty.$$

Then there is an $\mathfrak{L}^2$ *kernel* K *such that*

$$K = {}^\circ\sum_{n=1}^{\infty} \kappa_n(e_n \otimes f_n) \quad (\mathfrak{L}^2), \tag{1}$$

and we have

$$(Kf_n, e_n) = \kappa_n \quad (n = 1, 2, \ldots), \tag{2}$$

$$\| K \|^2 = \sum_{n=1}^{\infty} |\kappa_n|^2. \tag{3}$$

Furthermore, if x *is an arbitrary* $\mathfrak{L}^2$ *function, then*

$$Kx = {}^\circ\sum_{n=1}^{\infty} \kappa_n(x, f_n)\,e_n \quad (\mathfrak{L}^2). \tag{4}$$

The series on the right-hand side of (1) is, by the Riesz–Fischer theorem, convergent in mean square to an $\mathfrak{L}^2$ function, which is defined only up to equivalence; as we showed in § 1·6, we can take it to be an $\mathfrak{L}^2$ kernel $K(s, t)$. By Theorem 4·3·2,

$$\begin{aligned}\kappa_n &= (K, e_n \otimes f_n) \\ &= \iint K(s, t)\,\overline{e_n(s)}\,f_n(t)\,ds\,dt \\ &= (Kf_n, e_n),\end{aligned}$$

which proves (2); (3) follows at once from § 4·3 (3). To obtain (4), we write

$$K_n = \sum_{\nu=1}^{n} \kappa_\nu (e_\nu \otimes f_\nu) \quad (n = 1, 2, \ldots).$$

Then
$$K_n x = \sum_{\nu=1}^{n} \kappa_\nu (x, f_\nu)\, e_\nu, \tag{5}$$

and, by Schwarz's inequality,

$$\begin{aligned} \| Kx - K_n x \|^2 &= \int ds \left| \int [K(s,t) - K_n(s,t)]\, x(t)\, dt \right|^2 \\ &\leqslant \int ds \int | K(s,t) - K_n(s,t) |^2 dt \int | x(u) |^2 du \\ &= \| K - K_n \|^2 \| x \|^2 \\ &\to 0 \quad (n \to \infty). \end{aligned}$$

This means, by (5), that

$$Kx = {}^\circ\sum_{n=1}^{\infty} \kappa_n (x, f_n)\, e_n \quad (\mathfrak{L}^2),$$

the required result.

CHAPTER V

THE CLASSICAL FREDHOLM THEORY

5·1. Introductory remarks. In Chapter III we proved Fredholm's determinant-free theorems for linear integral equations of the second kind. We shall now show how an explicit formula for the solution of such equations can be obtained, in terms of the expressions known as the Fredholm determinants. In dealing with this part of the subject we shall find it convenient to deal separately with continuous kernels and with $\mathfrak{L}^2$ kernels; the present chapter is devoted to the continuous case, and Chapter VI to the $\mathfrak{L}^2$ case.

We consider the integral equation

$$x(s) = y(s) + \lambda \int_a^b K(s,t)\, x(t)\, dt \quad (a \leqslant s \leqslant b), \tag{1}$$

where $K(s,t)$ is continuous in $a \leqslant s \leqslant b$, $a \leqslant t \leqslant b$, $y(s)$ is continuous in $a \leqslant s \leqslant b$; we seek solutions $x(s)$ that are continuous in the same interval.

The exposition of the theory given in this chapter is almost identical with Fredholm's (1903). The basic idea is to regard (1) as the limiting form of a finite system of linear equations in a finite number of variables, the kind of approximation used being rather different from that of Chapter III.

Let n be a positive integer, and let

$$a = s_0 < s_1 < s_2 < \dots < s_n = b$$

define a subdivision of (a, b) into n equal sub-intervals, each of length $\delta_n = (b-a)/n$; write

$$x(s_\nu) = x_\nu, \quad y(s_\nu) = y_\nu, \quad K(s_\mu, s_\nu) = k_{\mu\nu} \quad (\mu, \nu = 1, 2, \dots, n).$$

When $s = s_\mu$, (1) becomes

$$x_\mu = y_\mu + \lambda \int_a^b K(s_\mu, t)\, x(t)\, dt; \tag{2}$$

the right-hand side of (2) is, for large n, approximately equal to

$$y_\mu + \lambda \delta_n \sum_{\nu=1}^{n} k_{\mu\nu} x_\nu.$$

Thus (1) may be regarded as the limiting form, when $n \to \infty$, of the system of equations

$$x_\mu = y_\mu + \lambda\delta_n \sum_{\nu=1}^{n} k_{\mu\nu} x_\nu \quad (1 \leqslant \mu \leqslant n). \tag{3}$$

In vector notation, (3) becomes

$$\mathbf{x} = \mathbf{y} + \lambda\delta_n \mathbf{K}\mathbf{x}$$

or

$$(\mathbf{I} - \lambda\delta_n \mathbf{K})\mathbf{x} = \mathbf{y}. \tag{4}$$

The equation (4) has a unique solution $\mathbf{x}$ for each given $\mathbf{y}$ if and only if

$$d_n(\lambda) = \det(\mathbf{I} - \lambda\delta_n \mathbf{K}) \neq 0.$$

Writing the determinant explicitly, we have

$$d_n(\lambda) = \begin{vmatrix} 1-\lambda\delta_n k_{11} & -\lambda\delta_n k_{12} & \dots & -\lambda\delta_n k_{1n} \\ -\lambda\delta_n k_{21} & 1-\lambda\delta_n k_{22} & \dots & -\lambda\delta_n k_{2n} \\ \dots & \dots & \dots & \dots \\ -\lambda\delta_n k_{n1} & -\lambda\delta_n k_{n2} & \dots & 1-\lambda\delta_n k_{nn} \end{vmatrix}. \tag{5}$$

Expanding (5) as a polynomial in λ, we obtain

$$\begin{aligned} d_n(\lambda) = 1 - \lambda \sum_{\mu=1}^{n} \delta_n k_{\mu\mu} + \lambda^2 \sum_{1 \leqslant \mu < \nu \leqslant n} \delta_n^2 \begin{vmatrix} k_{\mu\mu} & k_{\mu\nu} \\ k_{\nu\mu} & k_{\nu\nu} \end{vmatrix} \\ - \lambda^3 \sum_{1 \leqslant \mu < \nu < \rho \leqslant n} \delta_n^3 \begin{vmatrix} k_{\mu\mu} & k_{\mu\nu} & k_{\mu\rho} \\ k_{\nu\mu} & k_{\nu\nu} & k_{\nu\rho} \\ k_{\rho\mu} & k_{\rho\nu} & k_{\rho\rho} \end{vmatrix} + \dots + (-1)^n \lambda^n \delta_n^n \det \mathbf{K} \\ = 1 - \lambda \sum_{\mu=1}^{n} \delta_n k_{\mu\mu} + \frac{\lambda^2}{2!} \sum_{\mu,\nu=1}^{n} \delta_n^2 \begin{vmatrix} k_{\mu\mu} & k_{\mu\nu} \\ k_{\nu\mu} & k_{\nu\nu} \end{vmatrix} \\ - \frac{\lambda^3}{3!} \sum_{\mu,\nu,\rho=1}^{n} \delta_n^3 \begin{vmatrix} k_{\mu\mu} & k_{\mu\nu} & k_{\mu\rho} \\ k_{\nu\mu} & k_{\nu\nu} & k_{\nu\rho} \\ k_{\rho\mu} & k_{\rho\nu} & k_{\rho\rho} \end{vmatrix} + \dots + (-1)^n \lambda^n \delta_n^n \det \mathbf{K}. \end{aligned} \tag{6}$$

If we now let $n \to \infty$, (6) becomes, at least formally,

$$d(\lambda) = \lim_{n\to\infty} d_n(\lambda)$$

$$= 1 - \lambda \int K(s,s)\,ds + \frac{\lambda^2}{2!}\iint \begin{vmatrix} K(s,s) & K(s,t) \\ K(t,s) & K(t,t) \end{vmatrix} ds\,dt$$

$$- \frac{\lambda^3}{3!}\iiint \begin{vmatrix} K(s,s) & K(s,t) & K(s,u) \\ K(t,s) & K(t,t) & K(t,u) \\ K(u,s) & K(u,t) & K(u,u) \end{vmatrix} ds\,dt\,du + \dots \quad (7)$$

If $d_n(\lambda) \neq 0$, (4) has the unique solution

$$\mathbf{x} = \frac{1}{d_n(\lambda)} \operatorname{adj}(\mathbf{I} - \lambda\delta_n \mathbf{K})\,\mathbf{y},$$

which we write in the form

$$x_\mu = \frac{1}{d_n(\lambda)} \sum_{\nu=1}^{n} \alpha_{\mu\nu}^{(n)} y_\nu$$

$$= \frac{1}{d_n(\lambda)} \left(\alpha_{\mu\mu}^{(n)} y_\mu + \delta_n \sum_{\nu \neq \mu} \frac{\alpha_{\mu\nu}^{(n)}}{\delta_n} y_\nu \right) \quad (1 \leqslant \mu \leqslant n).$$

Now $\alpha_{\mu\mu}^{(n)}$ is the determinant of the matrix derived from $\mathbf{I} - \lambda\delta_n\mathbf{K}$ by deleting the μth row and μth column, and is thus similar in structure to $d_n(\lambda)$; when $n \to \infty$, we therefore expect it to converge to $d(\lambda)$. When $\mu \neq \nu$, an argument similar to that which led to (6) gives

$$\alpha_{\mu\nu}^{(n)} = \lambda\delta_n \left\{ k_{\mu\nu} - \lambda \sum_{\rho=1}^{n} \delta_n \begin{vmatrix} k_{\mu\nu} & k_{\mu\rho} \\ k_{\rho\nu} & k_{\rho\rho} \end{vmatrix} \right.$$

$$\left. + \frac{\lambda^2}{2!} \sum_{\rho,\sigma=1}^{n} \delta_n^2 \begin{vmatrix} k_{\mu\nu} & k_{\mu\rho} & k_{\mu\sigma} \\ k_{\rho\nu} & k_{\rho\rho} & k_{\rho\sigma} \\ k_{\sigma\nu} & k_{\sigma\rho} & k_{\sigma\sigma} \end{vmatrix} - \dots \right\}. \quad (8)$$

When $n \to \infty$, $s_\mu \to s$, $s_\nu \to t$, we see from (8) that $\delta_n^{-1}\alpha_{\mu\nu}^{(n)}$ tends formally to $\lambda D_\lambda(s,t)$, where

$$D_\lambda(s,t) = K(s,t) - \lambda \int \begin{vmatrix} K(s,t) & K(s,u) \\ K(u,t) & K(u,u) \end{vmatrix} du$$

$$+ \frac{\lambda^2}{2!} \iint \begin{vmatrix} K(s,t) & K(s,u) & K(s,v) \\ K(u,t) & K(u,u) & K(u,v) \\ K(v,t) & K(v,u) & K(v,v) \end{vmatrix} du\,dv - \dots \quad (9)$$

These formal considerations suggest that, when $d(\lambda) \neq 0$, the solution of (1) may be given by the formula

$$x(s) = y(s) + \lambda \int_a^b \frac{D_\lambda(s,t)}{d(\lambda)} y(t)\,dt, \tag{10}$$

where $d(\lambda)$ and $D_\lambda(s,t)$ are defined by (7) and (9) respectively. This is in fact true, as we shall show; it is even possible to justify the above limiting process directly (Hilbert (1904)). It is easier, however, to do as Fredholm did; namely, begin with the expressions (7) and (9), prove the convergence of the infinite series involved in them, and then verify that (10) does give the solution of the original integral equation (1).

5·2. Hadamard's inequality. In proving the convergence of (7) and (9) we shall need the following result, which is usually known as Hadamard's inequality.†

THEOREM 5·2·1. *Let* $\mathbf{A} = [a_{\mu\nu}]$ *be an* $n \times n$ *matrix of complex numbers. Then*

$$|\det \mathbf{A}|^2 \leqslant \prod_{\mu=1}^{n} \sum_{\nu=1}^{n} |a_{\mu\nu}|^2. \tag{1}$$

Write
$$c_\mu^2 = \sum_{\nu=1}^{n} |a_{\mu\nu}|^2 \quad (1 \leqslant \mu \leqslant n).$$

If the $a_{\mu\nu}$ vary in such a way that all the c_μ remain constant, the point in Euclidean space of $2n^2$ dimensions whose coordinates are the real and imaginary parts of the $a_{\mu\nu}$ ranges over a bounded closed set E. Since $|\det \mathbf{A}|$ is a continuous function of these coordinates, its upper bound in the set E will be attained, say when

$$a_{\mu\nu} = v_{\mu\nu} \quad (\mu, \nu = 1, 2, \ldots, n).$$

Let $\mathbf{V} = [v_{\mu\nu}]$, and let $V_{\mu\nu}$ be the cofactor of $v_{\mu\nu}$ in $\det \mathbf{V}$; then

$$\det \mathbf{V} = \sum_{\nu=1}^{n} v_{\mu\nu} V_{\mu\nu} \quad (1 \leqslant \mu \leqslant n).$$

By Cauchy's inequality (§ 1·4),

$$|\det \mathbf{V}|^2 \leqslant \sum_{\nu=1}^{n} |v_{\mu\nu}|^2 \sum_{\nu=1}^{n} |V_{\mu\nu}|^2 = c_\mu^2 \sum_{\nu=1}^{n} |V_{\mu\nu}|^2. \tag{2}$$

† Hadamard (1893).

For a fixed value of μ, equality holds in (2) if and only if the sequences $(\bar{v}_{\mu\nu})_{1\leqslant\nu\leqslant n}$ and $(V_{\mu\nu})_{1\leqslant\nu\leqslant n}$ are proportional;† in other words, if and only if there are constants α_μ and β_μ, not both 0, such that

$$\alpha_\mu \bar{v}_{\mu\nu} = \beta_\mu V_{\mu\nu} \quad (1 \leqslant \nu \leqslant n). \tag{3}$$

We now remark that equality must hold in (2); for, if it does not, we can increase $|\det \mathbf{V}|$ without altering c_μ simply by making the sequence $(\bar{v}_{\mu\nu})$ proportional to $(V_{\mu\nu})$. The sequences $(\bar{v}_{\mu\nu})$ and $(V_{\mu\nu})$ are therefore proportional for each μ; consequently, if $\rho \neq \mu$,

$$\alpha_\mu \sum_{\nu=1}^{n} v_{\rho\nu}\bar{v}_{\mu\nu} = \beta_\mu \sum_{\nu=1}^{n} v_{\rho\nu} V_{\mu\nu} = 0, \tag{4}$$

by a standard result on determinants. If $\alpha_\mu = 0$, we see from (3) that $V_{\mu\nu} = 0$ $(1 \leqslant \nu \leqslant n)$, whence $\det \mathbf{V} = 0$, and there is nothing to prove. If, however, $\alpha_\mu \neq 0$, we have, writing $\mathbf{V}^*$ for the transposed complex conjugate of $\mathbf{V}$, and using (4),

$$|\det \mathbf{V}|^2 = \det \mathbf{V} . \overline{\det \mathbf{V}} = \det \mathbf{V} . \det \mathbf{V}^* = \det (\mathbf{V}\mathbf{V}^*)$$

$$= \begin{vmatrix} c_1^2 & 0 & \dots & 0 \\ 0 & c_2^2 & \dots & 0 \\ \dots & \dots & \dots & \dots \\ 0 & 0 & \dots & c_n^2 \end{vmatrix}$$

$$= c_1^2 c_2^2 \dots c_n^2.$$

Hence $$|\det \mathbf{A}|^2 \leqslant |\det \mathbf{V}|^2 = \prod_{\mu=1}^{n} c_\mu^2 = \prod_{\mu=1}^{n} \sum_{\nu=1}^{n} |a_{\mu\nu}|^2,$$

the required result.

Corollary. *If* $|a_{\mu\nu}| \leqslant M$ $(\mu, \nu = 1, 2, \dots, n)$, *then*

$$|\det \mathbf{A}|^2 \leqslant n^n M^{2n}.$$

When the numbers $a_{\mu\nu}$ are real, Hadamard's inequality has a simple geometrical meaning: of all n-dimensional parallelepipeds with given edge-lengths the rectangular one has the greatest volume.

† Hardy, Littlewood and Pólya (1934), p. 26.

5·3. The Fredholm expansions. We now investigate the convergence of the series defining $d(\lambda)$ and $D_\lambda(s,t)$. It is convenient to introduce the notation

$$K\begin{pmatrix} u_1, u_2, \ldots, u_n \\ v_1, v_2, \ldots, v_n \end{pmatrix} = \begin{vmatrix} K(u_1, v_1) & K(u_1, v_2) & \ldots & K(u_1, v_n) \\ K(u_2, v_1) & K(u_2, v_2) & \ldots & K(u_2, v_n) \\ \ldots & \ldots & \ldots & \ldots \\ K(u_n, v_1) & K(u_n, v_2) & \ldots & K(u_n, v_n) \end{vmatrix}.$$

The function $d(\lambda)$ is defined by the series

$$d(\lambda) = \sum_{n=0}^{\infty} d_n \lambda^n, \tag{1}$$

where $d_0 = 1$ and

$$d_n = \frac{(-1)^n}{n!} \int_a^b \int_a^b \ldots \int_a^b K\begin{pmatrix} u_1, u_2, \ldots, u_n \\ u_1, u_2, \ldots, u_n \end{pmatrix} du_1 du_2 \ldots du_n \quad (n \geqslant 1); \tag{2}$$

the kernel $D_\lambda(s,t)$ is defined by

$$D_\lambda(s,t) = \sum_{n=0}^{\infty} D_n(s,t) \lambda^n, \tag{3}$$

where $D_0(s,t) = K(s,t)$ and

$$D_n(s,t) = \frac{(-1)^n}{n!} \int_a^b \ldots \int_a^b K\begin{pmatrix} s, u_1, \ldots, u_n \\ t, u_1, \ldots, u_n \end{pmatrix} du_1 \ldots du_n \quad (n \geqslant 1). \tag{4}$$

The kernels $D_n(s,t)$ are clearly continuous.

THEOREM 5·3·1. *Let $K(s,t)$ be a continuous kernel, and define d_n and $D_n(s,t)$ as above. Then:* (a) *the series* (1) *is convergent for all complex λ, and $d(\lambda)$ is an integral function of λ;* (b) *for given (s,t), the series* (3) *is convergent for all complex λ, and $D_\lambda(s,t)$ is an integral function of λ;* (c) *the series* (3) *is uniformly absolutely convergent in (s,t,λ) when λ is confined to any bounded subset of the complex plane, and $D_\lambda(s,t)$ is a continuous kernel for every value of λ.*

We use the norm

$$\| K \| = \| K \|_c = (b-a) \sup_{(s,t)} | K(s,t) |$$

defined in § 1·5. By Theorem 5·2·1, Corollary, we have, if $|K(s,t)| \leqslant M$,

$$\left| K\begin{pmatrix} u_1, u_2, \ldots, u_n \\ v_1, v_2, \ldots, v_n \end{pmatrix} \right| \leqslant n^{\frac{1}{2}n} M^n;$$

hence, by (2),

$$|d_n| \leqslant \frac{n^{\frac{1}{2}n} \|K\|^n}{n!} = c_n \quad (n \geqslant 1), \tag{5}$$

say. Since

$$\frac{c_{n+1}}{c_n} = \frac{\|K\|}{\sqrt{(n+1)}} \left(1 + \frac{1}{n}\right)^{\frac{1}{2}n} \to 0 \ (n \to \infty),$$

the series $\Sigma c_n \lambda^n$ is convergent for all λ; *a fortiori* the same is true for $\Sigma d_n \lambda^n$, so that $d(\lambda)$ is an integral function of λ.

A similar application of Hadamard's inequality to (4) gives

$$(b-a) |D_{n-1}(s,t)| \leqslant \|D_{n-1}\| \leqslant \frac{n^{\frac{1}{2}n} \|K\|^n}{(n-1)!} \quad (n \geqslant 2), \tag{6}$$

from which the remaining statements of the theorem follow at once.

The function $d(\lambda)$ is called the *Fredholm determinant* of the kernel $K(s,t)$, and $D_\lambda(s,t)$ is its *first Fredholm minor*. Since $d_0 = 1$, $d(\lambda)$ does not vanish identically; since it is an integral function, its zeros form a finite or enumerable set,† say $\lambda_1, \lambda_2, \ldots$. We shall see in § 5·6 that these are precisely the characteristic values of $K(s,t)$.

5·4. The resolvent kernel. We shall now show that, when $d(\lambda) \neq 0$, λ is a regular value of $K(s,t)$, and that the resolvent kernel can then be expressed in terms of $d(\lambda)$ and $D_\lambda(s,t)$.

THEOREM 5·4·1. *Let $K(s,t)$ be a continuous kernel, and let $d(\lambda)$ and $D_\lambda(s,t)$ be its Fredholm determinant and first Fredholm minor. If $d(\lambda) \neq 0$, λ is a regular value of $K(s,t)$, and the resolvent kernel $H_\lambda(s,t)$ is given by*

$$H_\lambda(s,t) = \frac{D_\lambda(s,t)}{d(\lambda)}. \tag{1}$$

Equation (1) exhibits at once the meromorphic character of $H_\lambda(s,t)$ as a function of λ, a result that we obtained by a different method in Theorem 3·5·2.

We begin by proving the recurrence relation

$$D_n(s,t) = d_n K(s,t) + D_{n-1} K(s,t) \quad (n \geqslant 1). \tag{2}$$

† See, for example, Titchmarsh (1932), p. 246.

We have

$$D_n(s,t)=\frac{(-1)^n}{n!}$$

$$\times\int\cdots\int\begin{vmatrix} K(s,t) & K(s,u_1) & \dots & K(s,u_n) \\ K(u_1,t) & K(u_1,u_1) & \dots & K(u_1,u_n) \\ \dots & \dots & \dots & \dots \\ K(u_n,t) & K(u_n,u_1) & \dots & K(u_n,u_n) \end{vmatrix} du_1 \dots du_n$$

$$=d_n K(s,t)+Q_n(s,t), \tag{3}$$

where

$$Q_n(s,t)=\frac{(-1)^n}{n!}$$

$$\times\int\cdots\int\begin{vmatrix} 0 & K(s,u_1) & \dots & K(s,u_n) \\ K(u_1,t) & K(u_1,u_1) & \dots & K(u_1,u_n) \\ \dots & \dots & \dots & \dots \\ K(u_n,t) & K(u_n,u_1) & \dots & K(u_n,u_n) \end{vmatrix} du_1 \dots du_n. \tag{4}$$

We now expand the determinant in (4) by its first column, obtaining

$$Q_n(s,t)=\frac{(-1)^n}{n!}\sum_{\nu=1}^{n}(-1)^{\nu}\int\dots\int K\begin{pmatrix} s, u_1, \dots, u_{\nu-1}, u_{\nu+1}, \dots, u_n \\ u_1, u_2, \dots, u_{\nu}, u_{\nu+1}, \dots, u_n \end{pmatrix}$$

$$\times K(u_\nu, t)\, du_1 \dots du_n.$$

In the νth term on the right-hand side we replace the variables $u_\nu, u_{\nu+1}, \dots, u_n$ by $u, u_\nu, \dots, u_{n-1}$ respectively, and we bring the νth column of the determinant into the first place by $(\nu-1)$ interchanges of adjacent columns. We thus obtain

$$Q_n(s,t)$$

$$=\frac{(-1)^{n-1}}{n!}\sum_{\nu=1}^{n}\int\dots\iint K\begin{pmatrix} s, u_1, \dots, u_{n-1} \\ u, u_1, \dots, u_{n-1} \end{pmatrix} K(u,t)\, du_1 \dots du_{n-1}\, du$$

$$=\frac{(-1)^{n-1}}{(n-1)!}\int\dots\iint K\begin{pmatrix} s, u_1, \dots, u_{n-1} \\ u, u_1, \dots, u_{n-1} \end{pmatrix} K(u,t)\, du_1 \dots du_{n-1}\, du$$

$$=\int D_{n-1}(s,u)\, K(u,t)\, du$$

$$=D_{n-1} K(s,t). \tag{5}$$

The relation (2) now follows from (3) and (5).

We next multiply (2) by λ^n and sum over n; after adding $K(s,t)$ to both sides of the resulting equation, we have

$$D_\lambda(s,t) = d(\lambda)\,K(s,t) + \lambda D_\lambda K(s,t). \tag{6}$$

If we expand the determinant in (4) by its first row instead of its first column, a similar argument gives us

$$D_\lambda(s,t) = d(\lambda)\,K(s,t) + \lambda K D_\lambda(s,t). \tag{7}$$

When $d(\lambda) \neq 0$, we write

$$H_\lambda(s,t) = \frac{D_\lambda(s,t)}{d(\lambda)};$$

we then have, from (6) and (7),

$$H_\lambda(s,t) - K(s,t) = \lambda H_\lambda K(s,t) = \lambda K H_\lambda(s,t).$$

Thus $H_\lambda(s,t)$ satisfies the resolvent equation, and λ is a regular value of $K(s,t)$.

The following result now follows from Theorem 2·1·2.

THEOREM 5·4·2. *Let $K(s,t)$ be a continuous kernel, and $y(s)$ a continuous function. Let $d(\lambda)$ and $D_\lambda(s,t)$ be the Fredholm determinant and the first Fredholm minor of $K(s,t)$. If λ is not a zero of $d(\lambda)$, the integral equation*

$$x(s) = y(s) + \lambda \int_a^b K(s,t)\,x(t)\,dt \tag{8}$$

has the unique continuous solution

$$x(s) = y(s) + \frac{\lambda}{d(\lambda)} \int_a^b D_\lambda(s,t)\,y(t)\,dt.$$

In particular, the homgeneous equation

$$x(s) = \lambda \int_a^b K(s,t)\,x(t)\,dt$$

has the unique continuous solution $x(s) = 0$; every characteristic value of $K(s,t)$ must be a zero of $d(\lambda)$.

We have thus found an explicit formula for the solution of (8) in the case when λ is not a zero of the Fredholm determinant $d(\lambda)$; we shall see later that these values of λ exhaust the regular values of $K(s,t)$.

5·5. Recurrence relations. In the present section we collect some miscellaneous relations, some of which will be useful later.

THEOREM 5·5·1. *With the notations defined in* § 5·3, *we have*

$$D_n(s,t) = d_n K(s,t) + KD_{n-1}(s,t) = d_n K(s,t) + D_{n-1}K(s,t) \quad (n \geqslant 1), \quad (1)$$

$$d_n = -\frac{1}{n}\int D_{n-1}(s,s)\,ds \quad (n \geqslant 1), \quad (2)$$

$$d'(\lambda) = -\int D_\lambda(s,s)\,ds. \quad (3)$$

The second equation in (1) is the equation § 5·4 (2) and the first equation is proved similarly.

To obtain (2), we start from the equation

$$D_{n-1}(s,t) = \frac{(-1)^{n-1}}{(n-1)!}\int\ldots\int K\begin{pmatrix} s, u_1, \ldots, u_{n-1} \\ t, u_1, \ldots, u_{n-1}\end{pmatrix} du_1 \ldots du_{n-1} \quad (n \geqslant 1).$$

Putting $t = s$ and integrating, we have

$$\int D_{n-1}(s,s)\,ds = \frac{(-1)^{n-1}}{(n-1)!}\int\int\ldots\int K\begin{pmatrix} s, u_1, \ldots, u_{n-1} \\ s, u_1, \ldots, u_{n-1}\end{pmatrix} ds\,du_1 \ldots du_{n-1}$$

$$= -nd_n,$$

the required result. It follows at once that

$$d'(\lambda) = d_1 + 2d_2\lambda + 3d_3\lambda^2 + \ldots$$

$$= -\sum_{n=0}^{\infty} \lambda^n \int D_n(s,s)\,ds$$

$$= -\int D_\lambda(s,s)\,ds,$$

which is (3); the term-by-term integration is justified by using the inequality § 5·3 (6).

We remark that the coefficients d_n and $D_n(s,t)$ of the series for $d(\lambda)$ and $D_\lambda(s,t)$ can be calculated step by step from equations (1) and (2); starting with $d_0 = 1$, $D_0(s,t) = K(s,t)$, we find d_1 from (2) and $D_1(s,t)$ from (1), then d_2 from (2), $D_2(s,t)$ from (1), and so on. We can thus avoid the direct calculation of the rather unwieldy determinants in the explicit formulae for d_n and $D_n(s,t)$. This method is occasionally useful for kernels given by an explicit formula, but is not as a rule suitable for numerical work, since the series for $d(\lambda)$ and $D_\lambda(s,t)$ are usually rather slowly convergent except for small values of λ. Numerical methods for the solution of integral equations have recently been surveyed by Bückner (1952).

5·6. The homogeneous equation. We shall now show that every zero of the Fredholm determinant of a continuous kernel is actually a characteristic value of the kernel.

THEOREM 5·6·1. *Let $K(s,t)$ be a continuous kernel, and let $d(\lambda)$ be its Fredholm determinant. If λ_0 is a zero of $d(\lambda)$, the homogeneous equation*

$$x(s)=\lambda_0\int_a^b K(s,t)\,x(t)\,dt \tag{1}$$

has a continuous solution $x(s)$ that is not identically zero; in other words, λ_0 is a characteristic value of $K(s,t)$.

By § 5·4 (7), we have

$$D_\lambda(s,t)=d(\lambda)\,K(s,t)+\lambda\int K(s,u)\,D_\lambda(u,t)\,du. \tag{2}$$

When $\lambda=\lambda_0$, $d(\lambda_0)=0$, so that (2) becomes

$$D_{\lambda_0}(s,t)=\lambda_0\int K(s,u)\,D_\lambda(u,t)\,du. \tag{3}$$

If $D_{\lambda_0}(s,t)$ does not vanish identically as a function of (s,t), we can choose t_0 so that $D_{\lambda_0}(s,t_0)$ does not vanish identically as a function of s; writing

$$x(s)=D_{\lambda_0}(s,t_0),$$

we have

$$x(s)=\lambda_0\int K(s,u)\,x(u)\,du,$$

and $x(s)$ is the required characteristic function.

We remark that this case always applies if λ_0 is a simple zero of $d(\lambda)$, so that $d'(\lambda_0)\neq 0$; for, by Theorem 5·5·1,

$$\int D_{\lambda_0}(s,s)\,ds=-d'(\lambda_0)\neq 0,$$

so that $D_{\lambda_0}(s,t)$ cannot vanish identically.

If $D_{\lambda_0}(s,t)=0$ for all (s,t), we have to proceed a little differently. Since $D_\lambda(s,t)$ is an integral function of λ, we have

$$D_\lambda(s,t)=(\lambda-\lambda_0)^m C_m(s,t)+(\lambda-\lambda_0)^{m+1} C_{m+1}(s,t)+\dots \quad (4)$$

for some $m \geqslant 1$, where the coefficients $C_r(s,t)$ are given by

$$C_r(s,t)=\frac{1}{2\pi i}\int_\Gamma \frac{D_\lambda(s,t)}{(\lambda-\lambda_0)^{r+1}}\,d\lambda \quad (r \geqslant m), \quad (5)$$

Γ being a simple closed contour surrounding the point λ_0 once in the positive direction, and where $C_m(s,t)$ does not vanish identically. Since, by Theorem 5·3·1, $D_\lambda(s,t)$ is a continuous function of (s,t,λ), it follows from (5) that $C_r(s,t)$ is a continuous function of (s,t) for all $r \geqslant m$. Taking Γ to be a circle of centre λ_0 and radius R_0, we have

$$|C_r(s,t)| \leqslant \frac{M}{R_0^r},$$

where M is an upper bound of $|D_\lambda(s,t)|$ for λ on Γ and arbitrary (s,t); the series (4) is therefore uniformly absolutely convergent in (λ,s,t) for $|\lambda-\lambda_0| \leqslant R < R_0$. Since R_0 is arbitrary, we have uniform convergence when λ is confined to any bounded subset of the complex plane.

By Theorem 5·5·1,

$$d'(\lambda)=-\int D_\lambda(s,s)\,ds; \quad (6)$$

substituting (4) in the right-hand side of (6), we see that $d'(\lambda)$ has a zero of order at least m at λ_0, so that the zero λ_0 of $d(\lambda)$ must be of order at least $m+1$. We can therefore write

$$d(\lambda)=c_{m+1}(\lambda-\lambda_0)^{m+1}+c_{m+2}(\lambda-\lambda_0)^{m+2}+\dots. \quad (7)$$

Substituting the series (4) and (7) in equation (2), and equating coefficients of $(\lambda-\lambda_0)^m$, we obtain

$$C_m(s,t)=\lambda_0\int K(s,u)\,C_m(u,t)\,du. \quad (8)$$

Choosing t_0 so that $C_m(s,t_0)$ does not vanish identically as a function of s, and writing $x(s)=C_m(s,t_0)$, we see that $x(s)$ is a characteristic function belonging to the characteristic value λ_0.

An alternative proof of Theorem 5·6·1 can be obtained by using the higher Fredholm minors, which we have not defined here; see, for example, Lovitt (1924), pp. 46–55.

Theorems 5·4·2 and 5·6·1 together show that the zeros of the Fredholm determinant are precisely the characteristic values of the kernel; in other words, λ is a regular value if and only if it is not a zero of $d(\lambda)$. We saw in Theorems 3·6·1 that the maximum number of linearly independent characteristic functions belonging to a given characteristic value λ_0 is finite; it can be shown† that this number cannot exceed the multiplicity of λ_0 as a zero of $d(\lambda)$. The two numbers are not necessarily equal, however, as we can see from the following example. Let

$$K(s,t)=1+t+3st \quad (-1\leqslant s\leqslant 1,\ -1\leqslant t\leqslant 1).$$

A little calculation shows that

$$d(\lambda)=1-4\lambda+4\lambda^2,$$

which has a double zero when $\lambda=\frac{1}{2}$, and no other zeros. If $x(s)$ is a characteristic function, we see from the equation

$$\begin{aligned} x(s) &= \frac{1}{2}\int_{-1}^{1}(1+t+3st)\,x(t)\,dt \\ &= \frac{1}{2}\int_{-1}^{1}(1+t)\,x(t)\,dt+\tfrac{3}{2}s\int_{-1}^{1}tx(t)\,dt \end{aligned}$$

that $x(s)$ must be of the form $A+Bs$, where A and B are constants. Substituting this expression as a trial solution, we find

$$A+Bs=A+\tfrac{1}{3}B+Bs,$$

whence $B=0$; all the characteristic functions are multiples of $x(s)=1$, so we do not obtain two linearly independent characteristic functions. On the other hand, the kernel

$$K(s,t)=1+3st \quad (-1\leqslant s\leqslant 1,\ -1\leqslant t\leqslant 1)$$

has the same Fredholm determinant, but has two linearly independent characteristic functions $x_1(s)=1$ and $x_2(s)=s$. This phenomenon is analogous to the one encountered in matrix theory and dealt with by the theory of elementary divisors; a similar theory can be constructed in the present case, and an account of it is given by Goursat (1927), chapter 31.

† Lovitt (1924), p. 52.

5·7. Concluding remarks. The theory described in the present chapter cannot be applied without modification to a general $\mathfrak{L}^2$ kernel. The main reason for this is the occurrence of such expressions as

$$\int_a^b K(u,u)\,du \tag{1}$$

in the formulae for d_n and $D_n(s,t)$. For a general $\mathfrak{L}^2$ kernel $K(s,t)$, $K(u,u)$ is not necessarily even a measurable function of u, so that the integral (1) may well be meaningless. In Chapter VI we shall show how this difficulty can be overcome; at the same time we shall reformulate the theory in another way, which throws additional light on the structure of the Fredholm determinant and the first Fredholm minor.

CHAPTER VI

THE FREDHOLM FORMULAE FOR $\mathfrak{L}^2$ KERNELS

6·1. The trace. In §5·7 we remarked that, when $K(s,t)$ is an $\mathfrak{L}^2$ kernel, the expression

$$\int K(t,t)\,dt \tag{1}$$

does not necessarily have a meaning. If, however, $K(t,t)$ is measurable and summable, we shall call (1) the *trace* of K and denote it by $\tau(K)$. It will be an essential tool in our treatment of $\mathfrak{L}^2$ kernels.

THEOREM 6·1·1. *Let $K(s,t)$ be an $\mathfrak{L}^2$ kernel. If (a) K can be expressed in the form GH, where $G(s,t)$ and $H(s,t)$ are $\mathfrak{L}^2$ kernels, or (b) K is of finite rank, then $\tau(K)$ exists.*

To prove (*a*), we remark that

$$K(t,t) = \int G(t,u)\,H(u,t)\,du$$

for all t. Since, by Schwarz's inequality, the double integral

$$\iint |\, G(t,u)\,H(u,t)\,|\,dt\,du$$

exists, so does the repeated integral

$$\int dt \int G(t,u)\,H(u,t)\,du = \int K(t,t)\,dt;$$

hence, in this case,

$$\tau(K) = \iint G(t,u)\,H(u,t)\,dt\,du. \tag{2}$$

In case (*b*), we can write

$$K(s,t) = \sum_{\nu=1}^{n} a_\nu(s)\,\overline{b_\nu(t)},$$

so that

$$K(t,t) = \sum_{\nu=1}^{n} a_\nu(t)\,\overline{b_\nu(t)};$$

it follows at once that $K(t,t)$ is measurable and summable, and that

$$\tau(K)=\sum_{\nu=1}^{n}(a_\nu, b_\nu).$$

The most important properties of the trace are summarized in the following theorem.

THEOREM 6·1·2. *Let $K(s,t)$ and $L(s,t)$ be $\mathfrak{L}^2$ kernels. Then:*

(*a*) *if $\tau(K)$ and $\tau(L)$ exist, then $\tau(\alpha K+\beta L)$ exists for arbitrary scalars α and β, and $\tau(\alpha K+\beta L)=\alpha\tau(K)+\beta\tau(L)$;*

(*b*) *if $\tau(K)$ exists, so does $\tau(K^*)$, and $\tau(K^*)=\overline{\tau(K)}$;*

(*c*) $\tau(KL)=\tau(LK)$;

(*d*) $|\tau(KL)|\leqslant\|K\|.\|L\|$;

(*e*) *$\tau(K^n)$ exists for $n\geqslant 2$, and $|\tau(K^n)|\leqslant\|K\|^n$;*

(*f*) *if $\|K_n-K\|\to 0$ and $\|L_n-L\|\to 0$, then $\tau(K_nL_n)\to\tau(KL)$.*

Items (*a*) and (*b*) are obvious, (*c*) follows at once from (2), and (*d*) from (2) and Schwarz's inequality; (*e*) is a consequence of (*d*), for we have

$$|\tau(K^n)|=|\tau(K^{n-1}K)|\leqslant\|K^{n-1}\|.\|K\|\leqslant\|K\|^n \quad (n\geqslant 2).$$

To prove (*f*), we have, by (*a*) and (*d*),

$$\begin{aligned}&|\tau(K_nL_n)-\tau(KL)|\\ &\quad=|\tau[(K_n-K)(L_n-L)]+\tau[K(L_n-L)]+\tau[(K_n-K)L]|\\ &\quad\leqslant\|K_n-K\|.\|L_n-L\|+\|K\|.\|L_n-L\|+\|K_n-K\|.\|L\|\\ &\quad\to 0 \quad (n\to\infty).\end{aligned}$$

6·2. Canonical kernels of finite rank. Instead of the type of approximation used in Chapter V, we return in the present chapter to approximation in mean square by kernels of finite rank, but the approximating kernels will now be taken in a special form.

If an $\mathfrak{L}^2$ kernel of finite rank can be written in the form

$$K=\sum_{\mu,\nu=1}^{p}k_{\mu\nu}e_\mu\otimes e_\nu,$$

where $\mathbf{K}=[k_{\mu\nu}]$ is a square matrix, and $(e_1, e_2, \ldots, e_p)$ is a finite orthonormal system, we shall call K a *canonical kernel of finite rank*. If we write $\tau(\mathbf{K})$ for the ordinary algebraic trace $\sum_\nu k_{\nu\nu}$ of the matrix $\mathbf{K}$, we have

$$K(t,t)=\sum_{\mu,\nu=1}^{p} k_{\mu\nu}e_\mu(t)\overline{e_\nu(t)},$$

whence
$$\tau(K)=\sum_{\mu,\nu=1}^{p} k_{\mu\nu}(e_\mu, e_\nu)=\sum_{\nu=1}^{p} k_{\nu\nu}=\tau(\mathbf{K}),$$

since $(e_1, \ldots, e_p)$ is orthonormal. If we write $[k_{\mu\nu}^{(r)}]=\mathbf{K}^r$, we have, as can easily be verified,

$$K^r=\sum_{\mu,\nu=1}^{p} k_{\mu\nu}^{(r)}\, e_\mu \otimes e_\nu \quad (r \geqslant 2),$$

so that K^r is a canonical kernel, and

$$\tau(K^r)=\sum_{\nu=1}^{p} k_{\nu\nu}^{(r)}=\tau(\mathbf{K}^r) \quad (r \geqslant 2).$$

We also have

$$\begin{aligned}\|K\|^2 &= \iint \{\sum_{\mu,\nu} k_{\mu\nu}e_\mu(s)\overline{e_\nu(t)}\}\{\sum_{\rho,\sigma} \overline{k_{\rho\sigma}}\,\overline{e_\rho(s)}\,e_\sigma(t)\}\,ds\,dt \\ &= \sum_{\mu,\nu,\rho,\sigma} (e_\mu, e_\rho)(e_\sigma, e_\nu)\,k_{\mu\nu}\overline{k_{\rho\sigma}} \\ &= \sum_{\mu,\nu} k_{\mu\nu}\overline{k_{\mu\nu}}=\sum_{\mu,\nu=1}^{p} |k_{\mu\nu}|^2.\end{aligned}$$

THEOREM 6·2·1. *Let*

$$K=\sum_{\rho=1}^{n} a_\rho \otimes b_\rho$$

be an $\mathfrak{L}^2$ kernel of finite rank. Then there is a canonical kernel

$$K_0=\sum_{\mu,\nu=1}^{p} k_{\mu\nu}e_\mu \otimes e_\nu$$

such that

(*a*) $$K(s,t)=^\circ K_0(s,t),$$

(*b*) $$\tau(K^r)=\tau(K_0^r)=\tau(\mathbf{K}^r) \quad (r=1,2,\ldots).$$

We apply the orthogonalization process of § 4·4 to the system of functions $(a_1, a_2, \ldots, a_n, b_1, b_2, \ldots, b_n)$, obtaining† a linearly

† If all the functions a_ρ and b_ρ are null, the resulting orthonormal system is void; in this case we take $K_0(s,t)=0$.

equivalent orthonormal system $(e_1, e_2, \ldots, e_p)$. We can then write

$$\left.\begin{aligned} a_\rho &= {}^\circ\sum_{\mu=1}^{p} a_{\rho\mu} e_\mu, \\ b_\rho &= {}^\circ\sum_{\nu=1}^{p} b_{\rho\nu} e_\nu, \end{aligned}\right\} \quad (1 \leqslant \rho \leqslant n),$$

so that

$$K = \sum_{\rho=1}^{n} a_\rho \otimes b_\rho = {}^\circ\sum_{\rho=1}^{n} \sum_{\mu,\nu=1}^{p} a_{\rho\mu} \overline{b_{\rho\nu}} e_\mu \otimes e_\nu = \sum_{\mu,\nu=1}^{p} k_{\mu\nu} e_\mu \otimes e_\nu,$$

where $$k_{\mu\nu} = \sum_{\rho=1}^{n} a_{\rho\mu} \overline{b_{\rho\nu}} \quad (1 \leqslant \mu \leqslant p,\ 1 \leqslant \nu \leqslant p).$$

To prove (*b*) for $r = 1$, we remark that

$$K(t,t) = \sum_{\rho=1}^{n} a_\rho(t) \overline{b_\rho(t)} = {}^\circ \sum_{\mu,\nu=1}^{p} k_{\mu\nu} e_\mu(t) \overline{e_\nu(t)},$$

whence $\tau(K) = \tau(K_0)$. For $r = 2$, we have

$$\tau(K^2) = \iint K(s,t)\, K(t,s)\, ds\, dt = \iint K_0(s,t)\, K_0(t,s)\, ds\, dt = \tau(K_0^2),$$

and the result for $r > 2$ follows by induction.

It follows from this result and Theorem 3·3·1 that we can approximate an arbitary $\mathfrak{L}^2$ kernel in mean square by canonical kernels of finite rank.

Let us now denote the zeros of the characteristic polynomial

$$\phi(\kappa) = \det(\mathbf{K} - \kappa \mathbf{I})$$

of the matrix $\mathbf{K}$ by $\kappa_1, \kappa_2, \ldots, \kappa_p$, repeating them according to their multiplicities, so that

$$\phi(\kappa) = \prod_{\nu=1}^{p} (\kappa_\nu - \kappa).$$

We also write $\sigma_n = \tau(K^n) = \tau(\mathbf{K}^n)$.

THEOREM 6·2·2. *If* κ_ν $(1 \leqslant \nu \leqslant p)$ *and* σ_n $(n \geqslant 1)$ *are defined as above, then*

$$\sigma_n = \sum_{\nu=1}^{p} \kappa_\nu^n \quad (n \geqslant 1).$$

We require the following lemma, which is well known in matrix theory; we give the proof for the sake of completeness.

LEMMA. *If*
$$\phi(\kappa)=\prod_{\nu=1}^{p}(\kappa_\nu-\kappa)$$
is the characteristic polynomial of the square matrix $\mathbf{K}$, *and* $g(t)$ *is an arbitrary polynomial with complex coefficients, then* $g(\mathbf{K})$ *has the characteristic polynomial*
$$\prod_{\nu=1}^{p}[g(\kappa_\nu)-\kappa].$$

Write
$$h(t)=g(t)-\kappa;$$
then, for some α,
$$h(t)=\alpha\prod_{\rho=1}^{r}(t-t_\rho),$$
where $t_1, t_2, \ldots, t_r$ are the zeros of $h(t)$, repeated according to their multiplicities. We then have
$$h(\mathbf{K})=\alpha\prod_{\rho=1}^{r}(\mathbf{K}-t_\rho\mathbf{I}).$$

Hence
$$\begin{aligned}\det[h(\mathbf{K})]&=\alpha^p\prod_{\rho=1}^{r}\det(\mathbf{K}-t_\rho\mathbf{I})\\&=\alpha^p\prod_{\rho=1}^{r}\phi(t_\rho)\\&=\alpha^p\prod_{\rho=1}^{r}\prod_{\nu=1}^{p}(\kappa_\nu-t_\rho)\\&=\prod_{\nu=1}^{p}\left\{\alpha\prod_{\rho=1}^{r}(\kappa_\nu-t_\rho)\right\}\\&=\prod_{\nu=1}^{p}h(\kappa_\nu),\end{aligned}$$
i.e.
$$\det[g(\mathbf{K})-\kappa\mathbf{I}]=\prod_{\nu=1}^{p}[g(\kappa_\nu)-\kappa],$$
the required result.

To prove the theorem, we take $g(t)=t^n$; the Lemma then gives
$$\det(\mathbf{K}^n-\kappa\mathbf{I})=\prod_{\nu=1}^{p}(\kappa_\nu^n-\kappa). \tag{3}$$
Expressing the two sides of (3) as polynomials in κ, and equating coefficients of κ^{p-1}, we have
$$\sigma_n=\sum_{\nu=1}^{p}\kappa_\nu^n,$$
as required.

6·3. The Fredholm formulae for canonical kernels of finite rank. We now consider the integral equation

$$x = y + \lambda K x, \tag{1}$$

where
$$K = \sum_{\mu,\nu=1}^{p} k_{\mu\nu} e_\mu \otimes e_\nu$$

is a canonical kernel of finite rank and y is an $\mathfrak{L}^2$ function. Writing

$$x_\mu = (x, e_\mu), \quad y_\mu = (y, e_\mu) \quad (1 \leqslant \mu \leqslant p),$$

we obtain from (1)

$$\begin{aligned} x_\rho &= y_\rho + \lambda (Kx, e_\rho) \\ &= y_\rho + \lambda \sum_{\mu,\nu=1}^{p} k_{\mu\nu} x_\nu (e_\mu, e_\rho) \\ &= y_\rho + \lambda \sum_{\nu=1}^{p} k_{\rho\nu} x_\nu \quad (1 \leqslant \rho \leqslant p). \end{aligned} \tag{2}$$

Conversely, if (x_ρ) is a solution of (2), and we take

$$x = y + \lambda \sum_{\mu,\nu=1}^{p} k_{\mu\nu} x_\nu e_\mu, \tag{3}$$

then x satisfies (1); for we have

$$\begin{aligned} (x, e_\rho) &= (y, e_\rho) + \lambda \sum_{\mu,\nu=1}^{p} k_{\mu\nu} x_\nu (e_\mu, e_\rho) \\ &= y_\rho + \lambda \sum_{\nu=1}^{p} k_{\rho\nu} x_\nu \\ &= x_\rho \quad (1 \leqslant \rho \leqslant p), \end{aligned}$$

so that (3) becomes

$$\begin{aligned} x &= y + \lambda \sum_{\mu,\nu=1}^{p} k_{\mu\nu} (x, e_\nu) e_\mu \\ &= y + \lambda K x. \end{aligned}$$

In vector notation, (2) is $\mathbf{x} = \mathbf{y} + \lambda \mathbf{K} \mathbf{x}$, or

$$(\mathbf{I} - \lambda \mathbf{K}) \mathbf{x} = \mathbf{y}. \tag{4}$$

If λ is not the reciprocal of a zero of the characteristic polynomial $\phi(\kappa) = \det(\mathbf{K} - \kappa \mathbf{I})$ of $\mathbf{K}$, equation (4) has a unique solution given by

$$\mathbf{x} = [d(\lambda)]^{-1} \mathbf{A}_\lambda \mathbf{y}, \tag{5}$$

where $\mathbf{A}_\lambda = \operatorname{adj}(\mathbf{I} - \lambda\mathbf{K})$, $d(\lambda) = \det(\mathbf{I} - \lambda\mathbf{K})$. Writing $\mathbf{A}_\lambda = [a_{\mu\nu}]$, we see that the corresponding solution of the integral equation (1) is

$$x = y + \frac{\lambda}{d(\lambda)} \sum_{\mu,\nu=1}^{p} k_{\mu\nu} \left(\sum_{\rho=1}^{p} a_{\nu\rho} y_\rho \right) e_\mu$$

$$= y + \frac{\lambda}{d(\lambda)} \sum_{\mu,\nu,\rho=1}^{p} k_{\mu\nu} a_{\nu\rho} (y, e_\rho)\, e_\mu$$

$$= y + \frac{\lambda}{d(\lambda)} D_\lambda y,$$

where
$$D_\lambda = \sum_{\mu,\nu,\rho=1}^{p} k_{\mu\nu} a_{\nu\rho} (e_\mu \otimes e_\rho) = K A_\lambda,$$

A_λ being the canonical kernel

$$A_\lambda = \sum_{\nu,\rho=1}^{p} a_{\nu\rho} (e_\nu \otimes e_\rho).$$

We are thus led to state the following theorem.

THEOREM 6·3·1. *Let*

$$K = \sum_{\mu,\nu=1}^{p} k_{\mu\nu} e_\mu \otimes e_\nu$$

be a canonical kernel, and suppose that $d(\lambda) = \det(\mathbf{I} - \lambda\mathbf{K}) \neq 0$. *Write*

$$\operatorname{adj}(\mathbf{I} - \lambda\mathbf{K}) = \mathbf{A}_\lambda = [a_{\mu\nu}],$$

$$A_\lambda = \sum_{\mu,\nu=1}^{p} a_{\mu\nu} (e_\mu \otimes e_\nu), \quad D_\lambda = K A_\lambda.$$

Then K has the resolvent

$$H_\lambda = [d(\lambda)]^{-1} D_\lambda.$$

By the definition of $\mathbf{A}_\lambda$,

$$\lambda \mathbf{K} \mathbf{A}_\lambda = \mathbf{A}_\lambda - (\mathbf{I} - \lambda\mathbf{K})\, \mathbf{A}_\lambda = \mathbf{A}_\lambda - d(\lambda)\, \mathbf{I}.$$

Similarly
$$\lambda \mathbf{A}_\lambda \mathbf{K} = \mathbf{A}_\lambda - d(\lambda)\, \mathbf{I},$$

so $\mathbf{A}_\lambda \mathbf{K} = \mathbf{K}\mathbf{A}_\lambda = \mathbf{D}_\lambda$, say. Hence

$$\lambda \mathbf{K} \mathbf{D}_\lambda = \mathbf{K}[\mathbf{A}_\lambda - d(\lambda)\,\mathbf{I}] = \mathbf{D}_\lambda - d(\lambda)\, \mathbf{K};$$

similarly
$$\lambda \mathbf{D}_\lambda \mathbf{K} = \mathbf{D}_\lambda - d(\lambda)\, \mathbf{K}.$$

Since D_λ is a canonical kernel with matrix $\mathbf{KA}_\lambda = \mathbf{D}_\lambda$, it follows at once that

$$\lambda K D_\lambda = \lambda D_\lambda K = D_\lambda - d(\lambda) K,$$

so that, provided $d(\lambda) \neq 0$, $H_\lambda = [d(\lambda)]^{-1} D_\lambda$ satisfies the resolvent equation.

Our next task is to find explicit expressions for $d(\lambda)$ and D_λ. Both of them are polynomials in λ; it is more convenient, however, to write them as infinite power series

$$\left.\begin{aligned} d(\lambda) &= \sum_{n=0}^{\infty} d_n \lambda^n, \\ D_\lambda &= \sum_{n=0}^{\infty} D_n \lambda^n; \end{aligned}\right\} \tag{6}$$

the coefficients D_n are all canonical kernels.

THEOREM 6·3·2. *If*

$$K = \sum_{\mu, \nu = 1}^{p} k_{\mu\nu} e_\mu \otimes e_\nu$$

is a canonical kernel and

$$\det(\mathbf{I} - \lambda \mathbf{K}) = d(\lambda) = \sum_{n=0}^{\infty} d_n \lambda^n,$$

then $d_0 = 1$ *and*

$$d_n = \frac{(-1)^n}{n!} \begin{vmatrix} \sigma_1 & n-1 & 0 & \dots & 0 & 0 \\ \sigma_2 & \sigma_1 & n-2 & \dots & 0 & 0 \\ \dots & \dots & \dots & \dots & \dots & \dots \\ \sigma_{n-1} & \sigma_{n-2} & \sigma_{n-3} & \dots & \sigma_1 & 1 \\ \sigma_n & \sigma_{n-1} & \sigma_{n-2} & \dots & \sigma_2 & \sigma_1 \end{vmatrix} \quad (n \geqslant 1),$$

where $\sigma_n = \tau(K^n) \quad (n \geqslant 1).$

We have

$$d(\lambda) = \det(\mathbf{I} - \lambda \mathbf{K}) = \prod_{\nu=1}^{p} (1 - \kappa_\nu \lambda),$$

where the κ_ν are the zeros of the characteristic polynomial of $\mathbf{K}$; whence, if $|\lambda|$ is sufficiently small,

$$\frac{d'(\lambda)}{d(\lambda)} = -\sum_{\nu=1}^{p} \frac{\kappa_\nu}{1 - \kappa_\nu \lambda} = -\sum_{\nu=1}^{p} \sum_{n=0}^{\infty} \kappa_\nu^{n+1} \lambda^n = -\sum_{n=0}^{\infty} \lambda^n \sum_{\nu=1}^{p} \kappa_\nu^{n+1}.$$

Consequently, by Theorem 6·2·2,

$$\frac{d'(\lambda)}{d(\lambda)} = -\sum_{n=0}^{\infty} \sigma_{n+1} \lambda^n. \tag{7}$$

Since

$$d(\lambda) = \sum_{n=0}^{\infty} d_n \lambda^n, \tag{8}$$

we also have

$$d'(\lambda) = \sum_{n=0}^{\infty} (n+1)\, d_{n+1} \lambda^n. \tag{9}$$

Combining (7), (8) and (9), we obtain

$$\sum_{n=0}^{\infty} (n+1)\, d_{n+1} \lambda^n = -\sum_{r=0}^{\infty} d_r \lambda^r . \sum_{s=0}^{\infty} \sigma_{s+1} \lambda^s. \tag{10}$$

Equating the coefficients of λ^m on the two sides of (10) for $0 \leqslant m \leqslant n$, we obtain the system of equations

$$\left.\begin{array}{ll} d_1 & = -\sigma_1 d_0, \\ \sigma_1 d_1 + 2d_2 & = -\sigma_2 d_0, \\ \dots\dots\dots\dots\dots\dots\dots\dots & \\ \sigma_{n-1} d_1 + \sigma_{n-2} d_2 + \dots + n d_n & = -\sigma_n d_0. \end{array}\right\} \tag{11}$$

It is clear that $d_0 = 1$. We can then solve (11) for d_n by Cramer's rule, and this gives the required result after a little manipulation of the determinant in the numerator.

THEOREM 6·3·3. *If*

$$K = \sum_{\mu, \nu = 1}^{p} k_{\mu\nu} e_\mu \otimes e_\nu$$

is a canonical kernel, $d(\lambda) = \det(\mathbf{I} - \lambda \mathbf{K})$, *and* K *has the resolvent*

$$H_\lambda = [d(\lambda)]^{-1} D_\lambda,$$

where

$$D_\lambda = \sum_{n=0}^{\infty} D_n \lambda^n,$$

then $D_0 = K$, *and*

$$D_n = \frac{(-1)^n}{n!} \begin{vmatrix} K & n & 0 & 0 & \dots & 0 & 0 \\ K^2 & \sigma_1 & n-1 & 0 & \dots & 0 & 0 \\ K^3 & \sigma_2 & \sigma_1 & n-2 & \dots & 0 & 0 \\ \dots & \dots & \dots & \dots & \dots & \dots & \dots \\ K^n & \sigma_{n-1} & \sigma_{n-2} & \sigma_{n-3} & \dots & \sigma_1 & 1 \\ K^{n+1} & \sigma_n & \sigma_{n-1} & \sigma_{n-2} & \dots & \sigma_2 & \sigma_1 \end{vmatrix} \quad (n \geqslant 1) \tag{12}$$

where $\sigma_n = \tau(K^n)$ $(n \geqslant 1)$.

In the course of proving Theorem 6·3·1, we showed that

$$D_\lambda - d(\lambda)\,K = \lambda K D_\lambda. \tag{13}$$

Substituting the series (6) (which are actually finite) for $d(\lambda)$ and D_λ in (13), we obtain

$$\sum_{n=0}^{\infty} D_n \lambda^n - \sum_{n=0}^{\infty} d_n K \lambda^n = \sum_{n=0}^{\infty} K D_n \lambda^{n+1}; \tag{14}$$

equating coefficients of λ^n on the two sides of (14) for $n = 0, 1, 2, \ldots$, we get

$$\left.\begin{aligned} D_0 &= K, \\ D_n &= d_n K + K D_{n-1} \quad (n \geqslant 1). \end{aligned}\right\} \tag{15}$$

Let us denote the expression on the right-hand side of (12) by E_n for the moment, and take $E_0 = K$. If we expand the determinant in (12) by its first row, we obtain

$$E_n = d_n K - n\left(-\frac{1}{n}\right) K E_{n-1} = d_n K + K E_{n-1} \quad (n \geqslant 1). \tag{16}$$

Thus $E_0 = D_0$, and (15) and (16) show that the sequences (D_n) and (E_n) are defined recursively by the same system of equations. We therefore have $D_n = E_n$ for all n, as required.

6·4. Modification of the formulae. The discussion given in § 6·3 is satisfactory for the comparatively trivial case of canonical kernels of finite rank, but it cannot be carried over as it stands to general $\mathfrak{L}^2$ kernels, since it still involves $\sigma_1 = \tau(K)$, which may not exist in the general case. In order to obtain formulae that are capable of generalization, we make the following apparently trivial modification.

Let $K(s,t)$ be a canonical kernel, and define $d(\lambda)$ and $D_\lambda(s,t)$ by the series (6) of § 6·3. Write

$$\delta(\lambda) = e^{\sigma_1 \lambda}\, d(\lambda), \quad \Delta_\lambda(s,t) = e^{\sigma_1 \lambda}\, D_\lambda(s,t). \tag{1}$$

If $d(\lambda) \neq 0$, the resolvent kernel of $K(s,t)$ is

$$H_\lambda(s,t) = \frac{D_\lambda(s,t)}{d(\lambda)} = \frac{\Delta_\lambda(s,t)}{\delta(\lambda)}.$$

Since $d(\lambda)$ is a polynomial, $\delta(\lambda)$ is an integral function of λ, so that we can write

$$\delta(\lambda)=\sum_{n=0}^{\infty}\delta_n\lambda^n, \tag{2}$$

the series being convergent for all λ. Similarly, $\Delta_\lambda(s,t)$ is an integral function of λ for fixed (s,t), and we can write

$$\Delta_\lambda(s,t)=\sum_{n=0}^{\infty}\Delta_n(s,t)\,\lambda^n, \tag{3}$$

the series being convergent for all λ. The coefficients $\Delta_n(s,t)$ are canonical kernels of finite rank, and we see without difficulty that the series in (3) is relatively uniformly absolutely convergent in (s,t) for each λ; precise inequalities for the coefficients $\Delta_n(s,t)$ will be obtained later.

Our next task is to obtain explicit formulae for δ_n and $\Delta_n(s,t)$.

THEOREM 6·4·1. *The coefficients* δ_n *defined by* (1) *and* (2) *are given by* $\delta_0=1$,

$$\delta_n=\frac{(-1)^n}{n!}\begin{vmatrix} 0 & n-1 & 0 & \dots & 0 & 0 & 0 \\ \sigma_2 & 0 & n-2 & \dots & 0 & 0 & 0 \\ \sigma_3 & \sigma_2 & 0 & \dots & 0 & 0 & 0 \\ \dots & \dots & \dots & \dots & \dots & \dots & \dots \\ \sigma_{n-1} & \sigma_{n-2} & \sigma_{n-3} & \dots & \sigma_2 & 0 & 1 \\ \sigma_n & \sigma_{n-1} & \sigma_{n-2} & \dots & \sigma_3 & \sigma_2 & 0 \end{vmatrix} \quad (n\geqslant 1). \tag{4}$$

Since $\delta(\lambda)=e^{\sigma_1\lambda}d(\lambda)$, it follows from § 6·3 (7) that, when $|\lambda|$ is sufficiently small,

$$\frac{\delta'(\lambda)}{\delta(\lambda)}=\sigma_1+\frac{d'(\lambda)}{d(\lambda)}=-(\sigma_2\lambda+\sigma_3\lambda^2+\dots). \tag{5}$$

The expression on the right-hand side of (5) differs from that in § 6·3 (7) only by the absence of σ_1; if we now repeat the arguments used in the proof of Theorem 6·3·2, we see that δ_n is equal to the expression obtained by replacing σ_1 by 0 in that for d_n. This is precisely (4) above. Finally, $\delta_0=d_0=1$.

THEOREM 6·4·2. *The coefficients* $\Delta_n(s,t)$ *defined by* (1) *and* (3) *are given by* $\Delta_0 = K$,

$$\Delta_n = \frac{(-1)^n}{n!} \begin{vmatrix} K & n & 0 & 0 & \dots & 0 & 0 & 0 \\ K^2 & 0 & n-1 & 0 & \dots & 0 & 0 & 0 \\ K^3 & \sigma_2 & 0 & n-2 & \dots & 0 & 0 & 0 \\ \dots & \dots & \dots & \dots & \dots & \dots & \dots & \dots \\ K^n & \sigma_{n-1} & \sigma_{n-2} & \sigma_{n-3} & \dots & \sigma_2 & 0 & 1 \\ K^{n+1} & \sigma_n & \sigma_{n-1} & \sigma_{n-2} & \dots & \sigma_3 & \sigma_2 & 0 \end{vmatrix}$$

$$(n \geqslant 1). \quad (6)$$

We argue exactly as in the proof of Theorem 6·3·3; since the series corresponding to those appearing in § 6·3 (14) are all relatively uniformly absolutely convergent, the fact that they are now genuinely infinite causes no difficulty.

The new expressions (4) and (6) do not involve $\sigma_1 = \tau(K)$, and we can therefore hope to carry over our results to general $\mathfrak{L}^2$ kernels. In order to do this, we must first obtain inequalities for the coefficients δ_n and Δ_n; these will be required in proving the convergence of the series (2) and (3) in the general case. We first quote a result from complex function theory.

LEMMA.† *Let*
$$f(\lambda) = \sum_{n=0}^{\infty} a_n \lambda^n$$
be an integral function of $\lambda = re^{i\theta}$, *and write*
$$M(r) = \sup_{|\lambda|=r} |f(\lambda)| \quad (r > 0).$$
Then
$$|a_n| \leqslant r^{-n} M(r) \quad (n \geqslant 0,\ r > 0).$$

THEOREM 6·4·3. *Let*
$$K = \sum_{\mu,\nu=1}^{p} k_{\mu\nu} e_\mu \otimes e_\nu$$
be a canonical kernel, $d(\lambda) = \det(\mathbf{I} - \lambda\mathbf{K})$, *and*
$$\delta(\lambda) = \sum_{n=0}^{\infty} \delta_n \lambda^n = e^{\sigma_1 \lambda} d(\lambda).$$

† See, for example, Titchmarsh (1932), p. 84.

Then $$|\delta_n| \leqslant \frac{e^{\frac{1}{2}n} \| K \|^n}{n^{\frac{1}{2}n}} \quad (n \geqslant 1). \tag{7}$$

The inequality (7) holds also when $n=0$ if we adopt the convention that the right-hand side is then equal to 1.

We may suppose that $\mathbf{K} \neq \mathbf{O}$, so that

$$\| K \|^2 = \sum_{\mu, \nu=1}^{p} | k_{\mu\nu} |^2 > 0,$$

for otherwise we should have $\delta_n = 0$ for $n \geqslant 1$, and the result would be trivial.

We have

$$\delta(\lambda) = e^{\sigma_1 \lambda} d(\lambda) = \exp\left(\lambda \sum_{\nu=1}^{p} k_{\nu\nu}\right) \det(\mathbf{I} - \lambda \mathbf{K})$$

$$= \begin{vmatrix} e^{\lambda k_{11}}(1-\lambda k_{11}) & \dots & -\lambda e^{\lambda k_{11}} k_{1p} \\ \dots & \dots & \dots \\ -\lambda e^{\lambda k_{pp}} k_{p1} & \dots & e^{\lambda k_{pp}}(1-\lambda k_{pp}) \end{vmatrix}.$$

To this equation we apply Hadamard's inequality (Theorem 5·2·1), obtaining

$$| \delta(\lambda) |^2 \leqslant \{| e^{2\lambda k_{11}} | (| 1-\lambda k_{11} |^2 + | \lambda k_{12} |^2 + \dots + | \lambda k_{1p} |^2)\}$$
$$\times \dots \times \{| e^{2\lambda k_{pp}} | (| \lambda k_{p1} |^2 + | \lambda k_{p2} |^2 + \dots + | 1-\lambda k_{pp} |^2)\}$$
$$= \prod_{\mu=1}^{p} \left\{\exp[2\Re(\lambda k_{\mu\mu})] \left[1 - 2\Re(\lambda k_{\mu\mu}) + | \lambda |^2 \sum_{\nu=1}^{p} | k_{\mu\nu} |^2\right]\right\}.$$

Since $1 + a \leqslant e^a$ for all real a, it follows that

$$| \delta(\lambda) |^2 \leqslant \prod_{\mu=1}^{p} \exp\left\{2\Re(\lambda k_{\mu\mu}) - 2\Re(\lambda k_{\mu\mu}) + | \lambda |^2 \sum_{\nu=1}^{p} | k_{\mu\nu} |^2\right\}$$
$$= \prod_{\mu=1}^{p} \exp\left\{| \lambda |^2 \sum_{\nu=1}^{p} | k_{\mu\nu} |^2\right\} = \exp(| \lambda |^2 . \| K \|^2),$$

whence $$| \delta(\lambda) | \leqslant \exp(\tfrac{1}{2} | \lambda |^2 \| K \|^2). \tag{8}$$

Applying the Lemma to (8), we obtain

$$| \delta_n | \leqslant r^{-n} \exp(\tfrac{1}{2} r^2 \| K \|^2) \quad (n \geqslant 1, r > 0).$$

The number r is at our disposal; we take $r = n^{\frac{1}{2}}/\| K \|$, so obtaining

$$|\delta_n| \leqslant \frac{\| K \|^n}{n^{\frac{1}{2}n}} \exp(\tfrac{1}{2}n) = \frac{e^{\frac{1}{2}n} \| K \|^n}{n^{\frac{1}{2}n}},$$

the required result.

THEOREM 6·4·4. *Let*

$$K = \sum_{\mu, \nu=1}^{p} k_{\mu\nu} e_\mu \otimes e_\nu$$

be a canonical kernel, $\mathbf{A} = [a_{\mu\nu}] = \operatorname{adj}(\mathbf{I} - \lambda \mathbf{K})$,

$$A_\lambda = \sum_{\mu, \nu=1}^{p} a_{\mu\nu} e_\mu \otimes e_\nu,$$

$$D_\lambda(s,t) = KA_\lambda(s,t) = A_\lambda K(s,t),$$

$$\Delta_\lambda(s,t) = \sum_{n=0}^{\infty} \Delta_n(s,t) \lambda^n = e^{\sigma_1 \lambda} D_\lambda(s,t).$$

Then
$$\| \Delta_n \| \leqslant \frac{e^{\frac{1}{2}(n+1)} \| K \|^{n+1}}{n^{\frac{1}{2}n}} \quad (n \geqslant 1). \tag{9}$$

Let $x(t)$ and $y(t)$ be $\mathfrak{L}^2$ functions such that $\| x \| = \| y \| = 1$, and write

$$x_\nu = (x, e_\nu), \quad y_\nu = (y, e_\nu) \quad (1 \leqslant \nu \leqslant p).$$

Then
$$(A_\lambda y, x) = \sum_{\mu, \nu=1}^{p} a_{\mu\nu} \bar{x}_\mu y_\nu$$

$$= -\begin{vmatrix} 0 & \bar{x}_1 & \dots & \bar{x}_p \\ y_1 & 1 - \lambda k_{11} & \dots & -\lambda k_{1p} \\ \dots & \dots & \dots & \dots \\ y_p & -\lambda k_{p1} & \dots & 1 - \lambda k_{pp} \end{vmatrix}, \tag{10}$$

by a well-known property of bordered determinants.

It is convenient to write

$$B_\lambda(s,t) = \sum_{n=0}^{\infty} B_n(s,t) \lambda^n = e^{\sigma_1 \lambda} A_\lambda(s,t),$$

so that
$$\Delta_\lambda = e^{\sigma_1 \lambda} D_\lambda = e^{\sigma_1 \lambda} A_\lambda K = B_\lambda K, \quad \Delta_n = B_n K.$$

Multiplying both sides of (10) by $e^{\sigma_1\lambda}$, we obtain

$$(B_\lambda y, x) = -\begin{vmatrix} 0 & \bar{x}_1 & \dots & \bar{x}_p \\ e^{\lambda k_{11}} y_1 & e^{\lambda k_{11}}(1-\lambda k_{11}) & \dots & -\lambda e^{\lambda k_{11}} k_{1p} \\ \dots & \dots & \dots & \dots \\ e^{\lambda k_{pp}} y_p & -\lambda e^{\lambda k_{pp}} k_{p1} & \dots & e^{\lambda k_{pp}}(1-\lambda k_{pp}) \end{vmatrix}. \quad (11)$$

We now apply Hadamard's inequality (Theorem 5·2·1) to (11), and argue as in the proof of Theorem 6·4·3, using the assumption that $\| x \| = \| y \| = 1$; this gives us

$$| (B_\lambda y, x) | \leqslant \exp(\tfrac{1}{2} + \tfrac{1}{2} | \lambda |^2 . \| K \|^2).$$

Since $(B_\lambda y, x)$ is linear in y and anti-linear in x (§ 1·4), it follows that for arbitrary $\mathfrak{L}^2$ functions x and y, not necessarily normalized,

$$| (B_\lambda y, x) | \leqslant \exp(\tfrac{1}{2} + \tfrac{1}{2} | \lambda |^2 . \| K \|^2) \| x \| . \| y \| . \quad (12)$$

Applying the Lemma to the regular function $(B_\lambda y, x)$ of λ, we obtain

$$| (B_n y, x) | \leqslant \frac{\exp(\tfrac{1}{2} + \tfrac{1}{2} r^2 \| K \|^2)}{r^n} \| x \| . \| y \| \quad (n \geqslant 1,\ r > 0).$$

Taking $r = n^{\frac{1}{2}} / \| K \|$, we have

$$| (B_n y, x) | \leqslant \frac{\exp(\tfrac{1}{2} + \tfrac{1}{2} n) \| K \|^n}{n^{\frac{1}{2} n}} \| x \| . \| y \| \quad (n \geqslant 1). \quad (13)$$

We now specialize the functions x and y, beginning by taking $x = B_n y$; (13) then becomes

$$\| B_n y \|^2 \leqslant \frac{e^{\frac{1}{2}(n+1)} \| K \|^n}{n^{\frac{1}{2} n}} \| B_n y \| . \| y \|,$$

whence

$$\| B_n y \| \leqslant \frac{e^{\frac{1}{2}(n+1)} \| K \|^n}{n^{\frac{1}{2} n}} \| y \| \quad (n \geqslant 1). \quad (14)$$

We now take $y(u) = K(u, t)$ for any fixed value of t; squaring both sides of (14), we have

$$\int ds \left| \int B_n(s, u) K(u, t) \, du \right|^2 \leqslant \frac{e^{n+1} \| K \|^{2n}}{n^n} \int | K(u, t) |^2 du. \quad (15)$$

Finally, integrating with respect to t, and using the fact that $B_n K = \Delta_n$, we obtain

$$\iint |\Delta_n(s,t)|^2 \, ds\, dt \leqslant \frac{e^{n+1} \| K \|^{2n}}{n^n} \iint |K(u,t)|^2 \, du\, dt,$$

whence $$\| \Delta_n \| \leqslant \frac{e^{\frac{1}{2}(n+1)} \| K \|^{n+1}}{n^{\frac{1}{2}n}} \quad (n \geqslant 1),$$

the required result.

6·5. The Fredholm formulae for general $\mathfrak{L}^2$ kernels. We are now in a position to carry over our results to general $\mathfrak{L}^2$ kernels. We begin by defining and establishing the properties of the coefficients δ_n and $\Delta_n(s,t)$ for this case.

THEOREM 6·5·1. *Let $K(s,t)$ be an $\mathfrak{L}^2$ kernel. Write*

$$\sigma_n = \tau(K^n) \quad (n \geqslant 2).$$

Define δ_n and $\Delta_n(s,t)$ for $n \geqslant 0$ by the formulae $\delta_0 = 1$, $\Delta_0 = K$,

$$\delta_n = \frac{(-1)^n}{n!} \begin{vmatrix} 0 & n-1 & 0 & \dots & 0 & 0 & 0 \\ \sigma_2 & 0 & n-2 & \dots & 0 & 0 & 0 \\ \sigma_3 & \sigma_2 & 0 & \dots & 0 & 0 & 0 \\ \dots & \dots & \dots & \dots & \dots & \dots & \dots \\ \sigma_{n-1} & \sigma_{n-2} & \sigma_{n-3} & \dots & \sigma_2 & 0 & 1 \\ \sigma_n & \sigma_{n-1} & \sigma_{n-2} & \dots & \sigma_3 & \sigma_2 & 0 \end{vmatrix} \quad (n \geqslant 1),$$

$$\Delta_n = \frac{(-1)^n}{n!} \begin{vmatrix} K & n & 0 & 0 & \dots & 0 & 0 & 0 \\ K^2 & 0 & n-1 & 0 & \dots & 0 & 0 & 0 \\ K^3 & \sigma_2 & 0 & n-2 & \dots & 0 & 0 & 0 \\ \dots & \dots & \dots & \dots & \dots & \dots & \dots & \dots \\ K^n & \sigma_{n-1} & \sigma_{n-2} & \sigma_{n-3} & \dots & \sigma_2 & 0 & 1 \\ K^{n+1} & \sigma_n & \sigma_{n-1} & \sigma_{n-2} & \dots & \sigma_3 & \sigma_2 & 0 \end{vmatrix}$$

$$(n \geqslant 1).$$

Then the following recurrence relations and inequalities hold:

$$\Delta_{n+1} = \delta_{n+1} K + K\Delta_n = \delta_{n+1} K + \Delta_n K \quad (n \geqslant 0); \tag{1}$$

$$\tau(\Delta_n - \delta_n K) = -(n+1)\,\delta_{n+1} \qquad (n \geqslant 0); \tag{2}$$

$$|\,\delta_n\,| \leqslant \frac{e^{\frac{1}{2}n}\,\|\,K\,\|^n}{n^{\frac{1}{2}n}} \qquad (n \geqslant 1); \tag{3}$$

$$\|\,\Delta_n\,\| \leqslant \frac{e^{\frac{1}{2}(n+1)}\,\|\,K\,\|^{n+1}}{n^{\frac{1}{2}n}} \qquad (n \geqslant 1). \tag{4}$$

The relation (1) is obtained by expanding the determinant in the expression for Δ_{n+1} by its first row, and (2) by taking the trace of the determinant in the expression for Δ_n after removing the element K in the top left-hand corner.

To prove the inequalities, we proceed as follows. By Theorem 3·3·1, there is a sequence (K_p) of $\mathfrak{L}^2$ kernels of finite rank such that

$$\lim_{p\to\infty} \|\,K_p - K\,\| = 0, \tag{5}$$

and, by Theorem 6·2·1, we may suppose that the kernels K_p are canonical. We write

$$\sigma_{n,p} = \tau(K_p^n) \quad (p \geqslant 1,\ n \geqslant 2),$$

and construct the coefficients $\delta_{n,p}$ and $\Delta_{n,p}$ corresponding to each of the kernels K_p. From (5) it follows inductively that

$$\lim_{p\to\infty} \|\,K_p^n - K^n\,\| = 0 \quad (n \geqslant 1); \tag{6}$$

for

$$\begin{aligned}\|\,K_p^n - K^n\,\| &= \|\,(K_p^{n-1} - K^{n-1})(K_p - K) + (K_p^{n-1} - K^{n-1})K \\ &\qquad + K^{n-1}(K_p - K)\,\| \\ &\leqslant \|\,K_p^{n-1} - K^{n-1}\,\|\,.\,\|\,K_p - K\,\| + \|\,K_p^{n-1} - K^{n-1}\,\|\,.\,\|\,K\,\| \\ &\qquad + \|\,K^{n-1}\|\,.\,\|\,K_p - K\,\| \\ &\to 0 \quad (p \to \infty),\end{aligned}$$

by the inductive hypothesis. Hence, by Theorem 6·1·2 (*f*),

$$\lim_{p\to\infty} \sigma_{n,p} = \lim_{p\to\infty} \tau(K_p^n) = \tau(K^n) = \sigma_n \quad (n \geqslant 2),$$

so that
$$\lim_{p\to\infty} \delta_{n,p} = \delta_n \quad (n \geqslant 0).$$

Next, since Δ_n is a linear combination of powers of K with coefficients involving the numbers σ_n, and $\Delta_{n,p}$ depends on K_p and $\sigma_{n,p}$ in the same way, we have

$$\lim_{p\to\infty} \| \Delta_{n,p} - \Delta_n \| = 0 \quad (n \geqslant 0).$$

Finally,

$$\lim_{p\to\infty} \| K_p \| = \| K \|, \quad \lim_{p\to\infty} \| \Delta_{n,p} \| = \| \Delta_n \| \quad (n \geqslant 0).$$

By Theorems 6·4·3 and 6·4·4, we have

$$| \delta_{n,p} | \leqslant \frac{e^{\frac{1}{2}n} \| K_p \|^n}{n^{\frac{1}{2}n}}, \quad \| \Delta_{n,p} \| \leqslant \frac{e^{\frac{1}{2}(n+1)} \| K_p \|^{n+1}}{n^{\frac{1}{2}n}} \quad (n \geqslant 1), \quad (7)$$

and the result now follows by letting $p \to \infty$ in (7).

THEOREM 6·5·2. *Let $K(s,t)$ be an $\mathfrak{L}^2$ kernel, and define δ_n and $\Delta_n(s,t)$ $(n \geqslant 0)$ as in Theorem 6·5·1. Write*

$$\delta(\lambda) = \sum_{n=0}^{\infty} \delta_n \lambda^n, \quad (8)$$

$$\Delta_\lambda(s,t) = \sum_{n=0}^{\infty} \Delta_n(s,t) \lambda^n, \quad (9)$$

Then the series in (8) is convergent for all complex λ, and $\delta(\lambda)$ is an integral function of λ; the series in (9) is relatively uniformly absolutely convergent in (s,t), $\Delta_\lambda(s,t)$ is an $\mathfrak{L}^2$ kernel for every value of λ, and $\Delta_\lambda(s,t)$ is an integral function of λ for fixed (s,t).

The convergence of (8) follows at once from the inequalities (3). It is clear from (4) that (9) is convergent in mean square for all λ; to prove relatively uniform convergence we use the recurrence relation (1). We have, in fact,

$$\begin{aligned} \Delta_{n+1} &= \delta_{n+1} K + \Delta_n K \\ &= \delta_{n+1} K + \delta_n K^2 + K \Delta_{n-1} K \quad (n \geqslant 1). \end{aligned}$$

By Theorem 2·1·3,

$$| K \Delta_{n-1} K(s,t) | \leqslant \| \Delta_{n-1} \| k_1(s) k_2(t),$$

where $$k_1(s) = \left\{ \int | K(s,u) |^2 du \right\}^{\frac{1}{2}}, \quad k_2(t) = \left\{ \int | K(u,t) |^2 du \right\}^{\frac{1}{2}}.$$

Hence

$$|\Delta_{n+1}(s,t)| \leqslant |\delta_{n+1}|.|K(s,t)| + |\delta_n|.|K^2(s,t)| + \|\Delta_{n-1}\|.k_1(s)k_2(t). \quad (10)$$

The relatively uniform absolute convergence of (9) is now a consequence of (3), (4) and (10), and the remaining assertions of the theorem follow at once.

We call $\delta(\lambda)$ the *modified Fredholm determinant* of $K(s,t)$ and $\Delta_\lambda(s,t)$ *the modified first Fredholm minor of* $K(s,t)$.

THEOREM 6·5·3. *Let $K(s,t)$ be an $\mathfrak{L}^2$ kernel, and let $\delta(\lambda)$ and $\Delta_\lambda(s,t)$ be its modified Fredholm determinant and first Fredholm minor. If $\delta(\lambda) \neq 0$, λ is a regular value of $K(s,t)$; and the resolvent $H_\lambda(s,t)$ is given by*

$$H_\lambda(s,t) = \frac{\Delta_\lambda(s,t)}{\delta(\lambda)}. \quad (11)$$

If $y(s)$ is an $\mathfrak{L}^2$ function, the integral equation

$$x(s) = y(s) + \lambda \int K(s,t)\, x(t)\, dt \quad (12)$$

has the unique $\mathfrak{L}^2$ solution

$$x(s) = y(s) + \lambda \int H_\lambda(s,t)\, y(t)\, dt = y(s) + \frac{\lambda}{\delta(\lambda)} \int \Delta_\lambda(s,t)\, y(t)\, dt.$$

By Theorem 6·5·1, $\Delta_0 = K$, and

$$\Delta_n = \delta_n K + \Delta_{n-1} K = \delta_n K + K\Delta_{n-1} \quad (n \geqslant 1).$$

Hence

$$\begin{aligned}\Delta_\lambda - \delta(\lambda) K &= \sum_{n=0}^{\infty} \Delta_n \lambda^n - \sum_{n=0}^{\infty} \delta_n \lambda^n K \\ &= \sum_{n=0}^{\infty} (\Delta_n - \delta_n K) \lambda^n \\ &= \sum_{n=0}^{\infty} \Delta_{n-1} K \lambda^n \\ &= \lambda \Delta_\lambda K.\end{aligned}$$

Similarly

$$\Delta_\lambda - \delta(\lambda) K = \lambda K \Delta_\lambda.$$

Thus, if $\delta(\lambda) \neq 0$, and H_λ is defined by (11), we have

$$H_\lambda - K = \lambda H_\lambda K = \lambda K H_\lambda,$$

so that H_λ is the resolvent of K. The final statement of the theorem now follows from Theorem 2·1·2.

It is interesting to note that the recurrence relation

$$\Delta_n = \delta_n K + \Delta_{n-1} K = \delta_n K + K\Delta_{n-1},$$

which is the only formal fact used in the proof of Theorem 6·5·3, was itself proved without using the definition of σ_n as the trace of K^n. The actual values of the constants σ_n are thus irrelevant to the truth of Theorem 6·5·3, provided only that the series defining $\delta(\lambda)$ and $\Delta_\lambda(s,t)$ converge; what we have actually proved is that there is a choice of values for the σ_n such that the series converge for all complex λ. The field of choice for the σ_n is not so wide as this remark might lead one to suppose; for, if the series for $\Delta_\lambda(s,t) = \delta(\lambda)\, H_\lambda(s,t)$ is to converge for all λ, the zeros of $\delta(\lambda)$ must cancel all the poles of the resolvent $H_\lambda(s,t)$, which depends only on $K(s,t)$ and not on the σ_n. Nevertheless, it can be easily proved, for instance, that any finite number of the σ_n can be replaced by 0, or by other constants chosen at random, without affecting the convergence of either series; we could even reintroduce σ_1 with an arbitrary value. On the other hand, if we were to take $\sigma_n = 0$ for all n, then $\delta(\lambda)$ would be identically equal to 1 and the series for $\Delta_\lambda(s,t)$ would be just the Neumann series for the resolvent, which is not in general convergent for all λ.

Expressions for the Fredholm determinant and its first minor in terms of the iterates of the kernel and their traces were first given by Plemelj (1904), and an abstract treatment of limited applicability was given by Michal and Martin (1934). The treatment given here is essentially that of Smithies (1941), which has been extended to general Banach spaces by Ruston (1951); see also Zaanen (1953).

6·6. Alternative formulae for δ_n and Δ_n. The coefficients δ_n and $\Delta_n(s,t)$ defined in Theorem 6·5·1 can be expressed in another form, which resembles much more closely the formulae for d_n and $D_n(s,t)$ given in § 5·3 for the case of continuous kernels.

THEOREM 6·6·1. *Let $K(s,t)$ be an $\mathfrak{L}^2$ kernel, and define δ_n and $\Delta_n(s,t)$ as in Theorem* 6·5·1. *Then*

$$\delta_n = \frac{(-1)^n}{n!}\int\cdots\int \begin{vmatrix} 0 & K(u_1,u_2) & \cdots & K(u_1,u_n) \\ K(u_2,u_1) & 0 & \cdots & K(u_2,u_n) \\ \cdots & \cdots & \cdots & \cdots \\ K(u_n,u_1) & K(u_n,u_2) & \cdots & 0 \end{vmatrix} \times du_1\ldots du_n \quad (n\geqslant 1), \tag{1}$$

$$\Delta_n(s,t) = \frac{(-1)^n}{n!}\int\cdots\int \begin{vmatrix} K(s,t) & K(s,u_1) & \cdots & K(s,u_n) \\ K(u_1,t) & 0 & \cdots & K(u_1,u_n) \\ \cdots & \cdots & \cdots & \cdots \\ K(u_n,t) & K(u_n,u_1) & \cdots & 0 \end{vmatrix} \times du_1\ldots du_n \quad (n\geqslant 1). \tag{2}$$

Denote the expressions on the right-hand sides of (1) and (2) by ϵ_n and $E_n(s,t)$ respectively. We clearly have

$$\epsilon_1 = 0 = \delta_1, \quad E_1(s,t) = K^2(s,t) = \Delta_1(s,t).$$

Also
$$\tau(E_n - \epsilon_n K) = -(n+1)\,\epsilon_{n+1} \quad (n\geqslant 1); \tag{3}$$

and finally, if we expand the determinant in (2) by its first row, and argue as in the proof of Theorem 5·4·1, we obtain

$$E_n = \epsilon_n K + KE_{n-1} = \epsilon_n K + E_{n-1}K \quad (n\geqslant 2). \tag{4}$$

The recurrence relations (3) and (4) are precisely those satisfied by (δ_n) and (Δ_n) (Theorem 6·5·1); since they define the sequences (ϵ_n) and (E_n) completely in terms of ϵ_1 and E_1, we have

$$\epsilon_n = \delta_n, \quad E_n = \Delta_n \quad (n\geqslant 1),$$

the required result.

The expressions (1) and (2) were given by Hilbert (1904) for a special type of discontinuous kernel; the first proof of their applicability to general $\mathfrak{L}^2$ kernels was given by Carleman (1921).

6·7. Characteristic values. Our next task is to show that every zero of the modified Fredholm determinant of an $\mathfrak{L}^2$ kernel $K(s,t)$ is a characteristic value of K. We begin with a lemma.

LEMMA. *Let $K(s,t)$ be an $\mathfrak{L}^2$ kernel, and λ_0 a zero of order M of its modified Fredholm determinant $\delta(\lambda)$. Then the resolvent $H_\lambda(s,t)$ can be written in the form*

$$H_\lambda(s,t) = \sum_{n=-M}^{\infty} A_n(s,t)\,(\lambda-\lambda_0)^n, \tag{1}$$

where the coefficients $A_n(s,t)$ are $\mathfrak{L}^2$ kernels, and the series is relatively uniformly absolutely convergent in (s,t) for all λ in some neighbourhood $|\lambda-\lambda_0|<r$ of the point λ_0.

Since $H_\lambda(s,t)=\Delta_\lambda(s,t)/\delta(\lambda)$ and, for given (s,t), $\Delta_\lambda(s,t)$ is regular at λ_0, an expansion of the form (1) certainly exists. Choose r so that λ_0 is the only zero of $\delta(\lambda)$ in $|\lambda-\lambda_0|\leqslant r$. Then $|\delta(\lambda)|^{-1}$ is bounded on the circumference $|\lambda-\lambda_0|=r$, and from the inequalities § 6·5 (10) it can easily be deduced that on the same circumference

$$|\Delta_\lambda(s,t)|\leqslant P(s,t),$$

where $P(s,t)$ is a non-negative $\mathfrak{L}^2$ kernel independent of λ. Hence

$$|H_\lambda(s,t)|\leqslant \beta P(s,t)$$

for some constant β. The coefficients $A_n(s,t)$ are given by

$$A_n(s,t)=\frac{1}{2\pi i}\int_C H_\lambda(s,t)\,(\lambda-\lambda_0)^{-n-1}\,d\lambda, \tag{2}$$

where C is the circumference $|\lambda-\lambda_0|=r$. Hence

$$|A_n(s,t)|\leqslant \beta r^{-n} P(s,t). \tag{3}$$

Since $A_n(s,t)$, as defined by (2), is clearly a measurable function of (s,t), of s for each value of t, and of t for each value of s, it follows from (3) that $A_n(s,t)$ is an $\mathfrak{L}^2$ kernel. Finally, by (3), the series (1) is relatively uniformly absolutely convergent in (s,t) for $|\lambda-\lambda_0|<r$.

THEOREM 6·7·1. *Let $K(s,t)$ be an $\mathfrak{L}^2$ kernel, and let λ_0 be a zero of its modified Fredholm determinant $\delta(\lambda)$. Then λ_0 is a characteristic value of $K(s,t)$.*

We begin by showing that terms of negative index are actually present in the expansion (1); in other words, that λ_0 is a genuine pole of the resolvent $H_\lambda(s,t)$ of $K(s,t)$. By Theorem 6·5·1, we have the recurrence relations

$$\tau(\Delta_n - \delta_n K) = -(n+1)\,\delta_{n+1} \quad (n \geqslant 0),$$
$$\Delta_n = \delta_n K + K\Delta_{n-1} \quad (n \geqslant 1).$$

Hence $\quad (n+1)\,\delta_{n+1} = -\tau(\Delta_n - \delta_n K) = -\tau(K\Delta_{n-1}) \quad (n \geqslant 1)$;
also $\delta_1 = 0$. Consequently

$$\delta'(\lambda) = \sum_{n=0}^{\infty} (n+1)\,\delta_{n+1}\lambda^n = -\sum_{n=1}^{\infty} \tau(K\Delta_{n-1})\,\lambda^n = -\lambda\tau(K\Delta_\lambda),$$

whence
$$\frac{\delta'(\lambda)}{\delta(\lambda)} = -\lambda\tau(KH_\lambda),$$

provided that $\delta(\lambda) \neq 0$. When $0 < |\lambda - \lambda_0| < r$, we therefore have, by (1),

$$\frac{\delta'(\lambda)}{\delta(\lambda)} = -\lambda \sum_{n=-M}^{\infty} \tau(KA_n)\,(\lambda - \lambda_0)^n. \tag{4}$$

Since $\delta'(\lambda)/\delta(\lambda)$ has a simple pole at λ_0, $\tau(KA_n) \neq 0$ for some $n < 0$, so that $A_n \neq 0$ for some $n < 0$. Let N be the greatest integer for which $A_{-N} \neq 0$.

When $0 < |\lambda - \lambda_0| < r$, the resolvent equation gives

$$H_\lambda - K = \lambda K H_\lambda = (\lambda - \lambda_0)\,KH_\lambda + \lambda_0 KH_\lambda. \tag{5}$$

Substituting the series (1) for H_λ in (5), and equating the coefficients of $(\lambda - \lambda_0)^{-N}$, we obtain

$$A_{-N} = \lambda_0 K A_{-N}. \tag{6}$$

Similarly
$$A_{-N} = \lambda_0 A_{-N} K, \tag{7}$$

whence
$$A_{-N} = \lambda_0^2 K A_{-N} K. \tag{8}$$

It follows from (8) that $\| A_{-N} \| \neq 0$; for, if $\| A_{-N} \| = 0$, $A_{-N}(s,t)$ would be a null function of (s,t), and (8) would then give $A_{-N} = 0$, contrary to hypothesis. We can therefore choose t_0 so that

$$\int |A_{-N}(s,t_0)|^2\,ds \neq 0;$$

if we then write $x(s) = A_{-N}(s,t_0)$, we shall have, by (6),

$$x(s) = \lambda_0 \int K(s,t)\,x(t)\,dt;$$

since $\| x \| \neq 0$, $x(s)$ is a characteristic function of $K(s,t)$ with characteristic value λ_0. This completes the proof.

COROLLARY. *If $K(s,t)$ is an $\mathfrak{L}^2$ kernel, every complex number λ is either a regular value or a characteristic value of K.*

For, if $\delta(\lambda) \neq 0$, λ is a regular value, by Theorem 6·5·3; and, if $\delta(\lambda) = 0$, λ is a characteristic value, by Theorem 6·7·1.

This result was obtained by quite a different method in Chapter III (Theorem 3·6·1). We shall now give direct proofs of the other determinant-free theorems of Chapter III; the alternative treatment given here is of some interest in its own right.

THEOREM 6·7·2. *Every characteristic value λ_0 of an $\mathfrak{L}^2$ kernel $K(s,t)$ is of finite rank.*

Let $x_1, x_2, \ldots, x_n$ be a finite linearly independent set of $\mathfrak{L}^2$ characteristic functions of $K(s,t)$ belonging to λ_0, so that

$$x_\nu = \lambda_0 K x_\nu \quad (1 \leqslant \nu \leqslant n). \tag{9}$$

It follows from (9) that the set $x_1, x_2, \ldots, x_n$ is also linearly independent with respect to equivalence; for, if

$$\sum_{\nu=1}^{n} \alpha_\nu x_\nu = {}^\circ 0,$$

then, by (9),

$$\sum_{\nu-1}^{n} \alpha_\nu x_\nu = \lambda_0 K\left(\sum_{\nu=1}^{n} \alpha_\nu x_\nu\right) = \lambda_0 K(0) = 0,$$

whence $\alpha_\nu = 0$ $(1 \leqslant \nu \leqslant n)$, by hypothesis. By the construction of Theorem 4·4·1, we can find an orthonormal set $y_1, y_2, \ldots, y_n$, containing exactly n functions, and such that every y_ν is a linear combination (in the ordinary sense) of the x_ν, and conversely. Every function y_ν is therefore an $\mathfrak{L}^2$ characteristic function of $K(s,t)$ belonging to λ_0, i.e.

$$y_\nu = \lambda_0 K y_\nu \quad (1 \leqslant \nu \leqslant n). \tag{10}$$

We now remark that $y_\mu \otimes y_\nu$ $(\mu, \nu = 1, 2, \ldots, n)$ is an orthonormal system of functions of two variables, and that

$$\begin{aligned}(K, y_\mu \otimes y_\nu) &= \iint K(s,t)\,\overline{y_\mu(s)}\,y_\nu(t)\,ds\,dt \\ &= \int \overline{y_\mu(s)}\,ds \int K(s,t)\,y_\nu(t)\,dt \\ &= (Ky_\nu, y_\mu) \\ &= \frac{(y_\nu, y_\mu)}{\lambda_0},\end{aligned}$$

by (10). Applying Bessel's inequality (Theorem 4·2·1) for functions of two variables, we obtain

$$\| K \|^2 \geqslant \sum_{\mu,\nu=1}^{n} |(K, y_\mu \otimes y_\nu)|^2 = \frac{1}{|\lambda_0|^2} \sum_{\mu,\nu=1}^{n} |(y_\nu, y_\mu)|^2 = \frac{n}{|\lambda_0|^2},$$

since (y_ν) is an orthonormal system. Thus $n \leqslant |\lambda_0|^2 . \| K \|^2$, and we have an upper bound for the number n.

The above argument gives us the following additional result.

COROLLARY. *Let λ_0 be a characteristic value of the $\mathfrak{L}^2$ kernel $K(s,t)$; then there is a finite orthonormal set $x_1, x_2, \ldots, x_n$ such that an $\mathfrak{L}^2$ function x is a characteristic function of $K(s,t)$ belonging to λ_0 if and only if x is a non-zero linear combination of $x_1, x_2, \ldots, x_n$.*

We may call $(x_1, x_2, \ldots, x_n)$ a *full orthonormal system* for (λ_0, K); the number n is equal to the rank of λ_0 as a characteristic value of K.

THEOREM 6·7·3. *If $K(s,t)$ is an $\mathfrak{L}^2$ kernel, and λ_0 is a characteristic value of $K(s,t)$, then $\bar{\lambda}_0$ is a characteristic value of $K^*(s,t) = \overline{K(t,s)}$, and the rank of λ_0 as a characteristic value of K is equal to the rank of $\bar{\lambda}_0$ as a characteristic value of K^*.*

Let $(x_1, x_2, \ldots, x_r)$ and $(y_1, y_2, \ldots, y_s)$ be full orthonormal systems for (λ_0, K) and $(\bar{\lambda}_0, K^*)$ respectively, and suppose, for instance, that $s > r$. Write

$$K_0 = K + \sum_{\rho=1}^{r} y_\rho \otimes x_\rho.$$

We shall show first that λ_0 is a regular value of K_0; to do this it is sufficient to show that it is not a characteristic value. Let x_0 be an $\mathfrak{L}^2$ function such that

$$x_0 = \lambda_0 K_0 x_0,$$

i.e.

$$x_0 - \lambda_0 K x_0 = \lambda_0 \sum_{\rho=1}^{r} (x_0, x_\rho) y_\rho. \tag{11}$$

Forming the inner product of (11) with y_σ, we obtain

$$(x_0 - \lambda_0 K x_0, y_\sigma) = \lambda_0 \sum_{\rho=1}^{r} (x_0, x_\rho)(y_\rho, y_\sigma) = \lambda_0 (x_0, x_\sigma) \quad (1 \leqslant \sigma \leqslant r). \tag{12}$$

On the other hand,

$$\begin{aligned}(x_0-\lambda_0 Kx_0, y_\sigma) &= (x_0, y_\sigma)-\lambda_0(Kx_0, y_\sigma)\\ &= (x_0, y_\sigma)-\lambda_0(x_0, K^*y_\sigma)\\ &= (x_0, y_\sigma-\bar{\lambda}_0 K^*y_\sigma)\\ &= 0. \qquad (13)\end{aligned}$$

By (12) and (13), $\quad (x_0, x_\sigma)=0 \quad (1\leqslant\sigma\leqslant r). \qquad (14)$

Hence, by (11), $x_0=\lambda_0 Kx_0$; since $(x_1, x_2, \ldots, x_r)$ is a full orthonormal system for (λ_0, K), there exist scalars $\alpha_1, \alpha_2, \ldots, \alpha_r$ such that

$$x_0=\sum_{\rho=1}^{r}\alpha_\rho x_\rho.$$

By (14), however, $\alpha_\rho=(x_0, x_\rho)=0$ $(1\leqslant\rho\leqslant r)$, so that $x_0=0$, and λ_0 is not a characteristic value of K_0.

Since λ_0 is a regular value of K_0, the equation

$$x=y_s+\lambda_0 K_0 x \qquad (15)$$

has an $\mathfrak{L}^2$ solution x. Equation (15) is equivalent to

$$x=y_s+\lambda_0 Kx+\lambda_0\sum_{\rho=1}^{r}(x, x_\rho)\,y_\rho. \qquad (16)$$

Forming the inner product of (16) with y_s, we obtain

$$(x, y_s)=1+(\lambda_0 Kx, y_s)=1+(x, \bar{\lambda}_0 K^*y_s)=1+(x, y_s),$$

whence $0=1$, a contradiction. The hypothesis $s>r$ is therefore untenable; a similar argument disposes of the hypothesis $s<r$, so we must have $s=r$, and the theorem is proved.

The last three theorems include the whole of Theorem 3·6·1. The result of Theorem 6·7·3 is the analogue for integral equations of the statement that a finite matrix and its transposed have the same rank. The next two theorems are also analogues of well-known results for finite matrices.

THEOREM 6·7·4. *If x is an $\mathfrak{L}^2$ characteristic function of the $\mathfrak{L}^2$ kernel K for the characteristic value λ, and y is an $\mathfrak{L}^2$ characteristic function of K^* for $\bar{\mu}$, where $\lambda\neq\mu$, then x and y are orthogonal.*

We have $\quad (x, y)=(\lambda Kx, y)=\lambda(Kx, y), \qquad (17)$

and also $$(x, y) = (x, \bar{\mu}K^*y) = \mu(Kx, y),$$

so that $\lambda(Kx, y) = \mu(Kx, y)$; since $\lambda \neq \mu$, $(Kx, y) = 0$, whence, by (17), $(x, y) = 0$.

THEOREM 6·7·5. *If K is an $\mathfrak{L}^2$ kernel, y is an $\mathfrak{L}^2$ function, and λ is a characteristic value of K, a necessary and sufficient condition for the integral equation*

$$x = y + \lambda Kx \tag{18}$$

to have an $\mathfrak{L}^2$ solution x is that y be orthogonal to all the $\mathfrak{L}^2$ characteristic functions of K^* belonging to the characteristic value $\bar{\lambda}$.

If (18) has an $\mathfrak{L}^2$ solution x, and u is an $\mathfrak{L}^2$ function such that $u = \bar{\lambda}K^*u$, then

$$(y, u) = (x - \lambda Kx, u) = (x, u) - \lambda(Kx, u) = (x, u) - (x, \bar{\lambda}K^*u) = 0.$$

The condition is therefore necessary.

Let us now suppose that y satisfies the conditions of the theorem. By the proof of Theorem 6·7·3, the equation

$$x = y + \lambda K_0 x = y + \lambda Kx + \lambda \sum_{\rho=1}^{r} (x, x_\rho)\, y_\rho \tag{19}$$

has a unique $\mathfrak{L}^2$ solution x. We then have

$$\begin{aligned} 0 &= (y, y_\sigma) \\ &= (x, y_\sigma) - (\lambda Kx, y_\sigma) - \lambda \sum_{\rho=1}^{r} (x, x_\rho)\,(y_\rho, y_\sigma) \\ &= (x, y_\sigma) - (x, \bar{\lambda}K^*y_\sigma) - \lambda(x, x_\sigma) \\ &= -\lambda(x, x_\sigma) \quad (1 \leqslant \sigma \leqslant r). \end{aligned}$$

Thus (19) is equivalent to $x = y + \lambda Kx$; in other words, x is an $\mathfrak{L}^2$ solution of (18). The condition is therefore sufficient.

The general solution of the integral equation (18) can now be obtained in the way described in Theorem 3·7·1.

CHAPTER VII

HERMITIAN KERNELS

7·1. Introductory remarks. The theory of linear integral equations of the second kind with Hermitian (in the real case, symmetric) $\mathfrak{L}^2$ kernels resembles in many respects and has close connexions with the theory of self-adjoint differential equations. We have illustrated these connexions in Chapter I; they are fully discussed and exploited in Goursat (1927), chapter XXXIII, Lovitt (1924) and Courant & Hilbert (1953), chapters V–VI.

Hilbert (1904) gave the first discussion of real symmetric kernels, using the theory of finite symmetric matrices as his starting point and then carrying through a limiting process. A direct treatment was given by E. Schmidt (1907 *a*); his methods require very little alteration to be applicable to symmetric $\mathfrak{L}^2$ kernels, and an outline of the necessary adaptations was given by Smithies (1937). The treatment given in the present chapter is based mainly on these, but has been extended to cover Hermitian as well as real symmetric kernels.

We saw in Chapter VI that the characteristic values of an $\mathfrak{L}^2$ kernel are precisely the zeros of its modified Fredholm determinant. It may happen that no such zeros exist, so that the kernel has no characteristic values and its resolvent is an integral function of the parameter λ; we have already seen in Chapter II that this happens in the case of Volterra kernels. An example of another kind is provided by the kernel

$$K(s,t) = \sin s \cos t \quad (0 \leqslant s \leqslant 2\pi,\ 0 \leqslant t \leqslant 2\pi).$$

Any solution of the homogeneous equation

$$x(s) = \lambda \int_0^{2\pi} \sin s \cos t\, x(t)\, dt = \lambda \sin s \int_0^{2\pi} x(t) \cos t\, dt \tag{1}$$

must clearly be of the form $x(s) = A \sin s$; substituting this in (1), we have

$$A \sin s = \lambda \sin s \int_0^{2\pi} A \sin t \cos t\, dt = 0,$$

whence $A = 0$, $x(s) = 0$. Thus this kernel has no characteristic values at all.

Our first task in the present chapter will be to show that the above phenomenon cannot occur for Hermitian $\mathfrak{L}^2$ kernels; more precisely, if a Hermitian $\mathfrak{L}^2$ kernel is not null, it possesses at least one characteristic value. From this it will follow that the kernel is determined (up to equivalence) by its characteristic values and the corresponding characteristic functions, and that the solutions of the corresponding non-homogeneous integral equation can be expressed in terms of the characteristic functions.

A large proportion of the present chapter is concerned with theorems on the expansion of the kernel and its iterates and of certain classes of $\mathfrak{L}^2$ functions in terms of the characteristic functions. These theorems are closely analogous to those concerning the transformation of quadratic forms in a finite number of variables into a sum of squares with real coefficients; they provide a canonical form for the linear operator defined by the kernel, and from this many of the properties of the kernel can be read off. The expansion theorems have numerous applications to problems concerning the expansion of arbitrary functions in terms of the eigenfunctions of a self-adjoint differential system; for these see, for example, Courant & Hilbert (1953), chapters V–VI.

7·2. The existence of characteristic values.

We shall give two proofs of the fundamental theorem on the existence of characteristic values. The first of these is essentially Kneser's (1906), and uses the Fredholm theory of Chapter VI; the second is an adaptation of Schmidt's (1907*a*) and is independent of the Fredholm theory.

THEOREM 7·2·1. *Every non-null Hermitian $\mathfrak{L}^2$ kernel $K(s,t)$ possesses at least one characteristic value.*

First proof. We require two lemmas.

LEMMA 1. *If $K(s,t)$ is a non-null Hermitian $\mathfrak{L}^2$ kernel, then all its iterates are non-null.*

Suppose that $K^n(s,t) =^\circ 0$ for some n. Choose r so that $2^r > n$; then

$$K^{2^r}(s,t) = K^n K^{2^r - n}(s,t) =^\circ 0.$$

Hence, for any $\mathfrak{L}^2$ function x, $(K^{2^r}x, x) = 0$; since K is Hermitian, it follows that

$$\| K^{2^{r-1}}x \|^2 = (K^{2^{r-1}}x, K^{2^{r-1}}x) = (K^{2^r}x, x) = 0,$$

whence $K^{2^{r-1}}x = ^\circ 0$. Consequently $(K^{2^{r-1}}x, y) = 0$ for arbitrary x and y. By choosing x and y appropriately, we deduce that

$$\int_{I_1} ds \int_{I_2} K^{2^{r-1}}(s,t)\, dt = 0$$

for arbitrary subintervals I_1 and I_2 of (a, b), whence it follows by standard Lebesgue integral theory that $K^{2^{r-1}}(s,t) = ^\circ 0$. By repeating the above argument we obtain $K^{2^{r-2}}(s,t) = ^\circ 0$, and so on, finally concluding that $K(s,t) = ^\circ 0$, contrary to hypothesis.

LEMMA 2. *If $K(s,t)$ is an $\mathfrak{L}^2$ kernel, $\sigma_n = \tau(K^n)$ $(n \geqslant 2)$ and $\delta(\lambda)$ is the modified Fredholm determinant of $K(s,t)$, then*

$$\frac{\delta'(\lambda)}{\delta(\lambda)} = -\sum_{n=1}^{\infty} \sigma_{n+1}\lambda^n, \tag{1}$$

provided that $|\lambda| < |\lambda_1|$, where λ_1 is the zero of $\delta(\lambda)$ of least absolute value: if $\delta(\lambda)$ has no zeros, (1) *holds for all complex λ.*

In the course of the proof of Theorem 6·7·1, we showed that

$$\frac{\delta'(\lambda)}{\delta(\lambda)} = -\lambda\tau(KH_\lambda),$$

where H_λ is the resolvent of K. By Theorem 2·5·1,

$$H_\lambda = K + \lambda K^2 + \lambda^2 K^3 + \ldots,$$

provided that $|\lambda|$ is sufficiently small, the series being relatively uniformly convergent. Hence

$$\frac{\delta'(\lambda)}{\delta(\lambda)} = -\lambda\tau\left(\sum_{n-1}^{\infty} \lambda^{n-1}K^{n+1}\right) = -\sum_{n=1}^{\infty} \sigma_{n+1}\lambda^n.$$

Thus (1) certainly holds when $|\lambda|$ is sufficiently small; the radius of convergence of the power series is clearly that given in the lemma.

We can now prove Theorem 7·2·1. Suppose that K has no characteristic values. Then $\delta(\lambda)$ has no zeros and by Lemma 2, (1) must hold for all λ. We shall show that this leads to a contradiction. The series on the right-hand side of (1) is absolutely convergent for all λ, whence the series

$$\sum_{n=1}^{\infty} u_n = \sum_{n=1}^{\infty} |\sigma_{2n}| \,.\, |\lambda|^{2n-1} \tag{2}$$

is convergent for all λ.

Since K is Hermitian, K^n is Hermitian for all n, and

$$\begin{aligned}\sigma_{2n} &= \tau(K^n K^n) \\ &= \iint K^n(s,t)\, K^n(t,s)\, ds\, dt \\ &= \iint K^n(s,t)\, \overline{K^n(s,t)}\, ds\, dt \\ &= \| K^n \|^2 \\ &> 0,\end{aligned} \tag{3}$$

by Lemma 1. Thus σ_{2n} is real and positive for all n. By Theorem 6·1·2,

$$\begin{aligned}\sigma_{2n}^2 &= [\tau(K^{n-1}K^{n+1})]^2 \\ &\leqslant \| K^{n-1} \|^2 \, \| K^{n+1} \|^2 \\ &= \sigma_{2n-2}\,\sigma_{2n+2},\end{aligned}$$

i.e.

$$\frac{\sigma_{2n+2}}{\sigma_{2n}} \geqslant \frac{\sigma_{2n}}{\sigma_{2n-2}} \quad (n \geqslant 2).$$

We therefore have

$$\frac{\sigma_{2n+2}}{\sigma_{2n}} \geqslant \frac{\sigma_{2n}}{\sigma_{2n-2}} \geqslant \dots \geqslant \frac{\sigma_4}{\sigma_2}.$$

Consequently, if u_n is defined by (2),

$$\frac{u_{n+1}}{u_n} = |\lambda|^2 \frac{\sigma_{2n+2}}{\sigma_{2n}} \geqslant |\lambda|^2 \frac{\sigma_4}{\sigma_2},$$

whence, by the ratio test for the convergence of series of positive terms, the series (2) is divergent when $|\lambda| > \surd(\sigma_2/\sigma_4)$, a contradiction. The kernel K must therefore possess at least one characteristic value; indeed, there must be one whose absolute value does not exceed $\sqrt{(\sigma_2/\sigma_4)}$.

Second proof. Instead of using the Fredholm theory as in the first proof, we can construct a characteristic function of the kernel $K(s,t)$ directly. We begin by proving two lemmas.

LEMMA 3. *Let* $\sigma_n = \tau(K^n)$ $(n \geqslant 2)$, *where* K *is a non-null Hermitian* $\mathfrak{L}^2$ *kernel. Then there is a real number* c *such that the limit*

$$C = \lim_{n\to\infty} \frac{\sigma_{2n}}{c^{2n}}$$

exists and is different from 0; *we then have* $C \geqslant 1$.

We write

$$\gamma_n = \frac{\sigma_{2n+2}}{\sigma_{2n}} \quad (n \geqslant 1),$$

and show as in the first proof that (γ_n) is non-decreasing. We also have, by (3),

$$\begin{aligned}\sigma_{2m+2n} &= \tau(K^{2m}K^{2n}) \\ &\leqslant \| K^{2m} \| . \| K^{2n} \| \\ &\leqslant \| K^m \|^2 \| K^n \|^2 \\ &= \sigma_{2m}\sigma_{2n},\end{aligned}$$

whence $$\sigma_{2m} \geqslant \frac{\sigma_{2m+2n}}{\sigma_{2n}} = \gamma_n \gamma_{n+1} \cdots \gamma_{m+n-1} \geqslant \gamma_n^m,$$

so that $$0 < \gamma_n \leqslant \sigma_{2m}^{1/m} \quad (m \geqslant 1,\ n \geqslant 1).$$

The sequence (γ_n) is thus bounded above; it is therefore convergent to a limit c^2, and we have

$$c^{2m} \leqslant \sigma_{2m} \quad (m \geqslant 1). \tag{4}$$

We now have $$\frac{\sigma_{2n+2}}{\sigma_{2n}} = \gamma_n \leqslant c^2,$$

which may be written $$\frac{\sigma_{2n+2}}{c^{2n+2}} \leqslant \frac{\sigma_{2n}}{c^{2n}};$$

the sequence $(c^{-2n}\sigma_{2n})$ is therefore non-increasing and, since, by (4), it is bounded below by 1, it has a limit $C \geqslant 1$.

LEMMA 4. *If K is an $\mathfrak{L}^2$ kernel and λ^2 is a characteristic value of K^2, then either λ or $-\lambda$ is a characteristic value of K.*

We note that in this lemma K is not required to be Hermitian.

Let

$$x = \lambda^2 K^2 x,$$

where $x \neq 0$. Write

$$x_1 = x + \lambda K x, \quad x_2 = x - \lambda K x.$$

Then
$$\lambda K x_1 = \lambda K x + \lambda^2 K^2 x = \lambda K x + x = x_1,$$

$$\lambda K x_2 = \lambda K x - \lambda^2 K^2 x = \lambda K x - x = -x_2.$$

Since $x = \tfrac{1}{2}x_1 + \tfrac{1}{2}x_2$, at least one of x_1 and x_2 is non-zero; thus either x_1 is a characteristic function of K with characteristic value λ, or x_2 is a characteristic function of K with characteristic value $-\lambda$ (or both).

We now come to the proof of the theorem. Write

$$G_n = c^{-2n} K^{2n} \quad (n \geqslant 1).$$

Then

$$\| G_m - G_n \|^2 = \iint [G_m(s,t) - G_n(s,t)]\,[\overline{G_m(s,t)} - \overline{G_n(s,t)}]\, ds\, dt$$

$$= \| G_m \|^2 - \tau(G_m G_n) - \tau(G_n G_m) + \| G_n \|^2$$

$$= c^{-4m}\sigma_{4m} - 2c^{-2m-2n}\sigma_{2m+2n} + c^{-4n}\sigma_{4n}$$

$$\to C - 2C + C = 0 \quad (m, n \to \infty),$$

by Lemma 3. The sequence (G_n) is therefore convergent in mean square to an $\mathfrak{L}^2$ kernel G, and we have

$$\| G \|^2 = \lim_{n\to\infty} \| G_n \|^2 = \lim_{n\to\infty} c^{-4n} \| K^{2n} \|^2 = \lim_{n\to\infty} c^{-4n}\sigma_{4n} = C \geqslant 1,$$

so that G is non-null. We also have

$$G_n = \frac{1}{c^2} K^2 G_{n-1} \quad (n > 1),$$

whence we obtain, by letting $n \to \infty$,

$$G(s,t) \overset{\circ}{=} \frac{1}{c^2} K^2 G(s,t) \overset{\circ}{=} \frac{1}{c^2} \int K^2(s,u)\, G(u,t)\, du. \tag{5}$$

We can now choose a value of t for which $x_0(s) = G(s,t)$ is a non-null $\mathfrak{L}^2$ function of s, and such that (5) holds for almost all s; we then have

$$x_0(s) = {}^{\circ}\frac{1}{c^2}\int K^2(s,u)\,x_0(u)\,du. \tag{6}$$

Writing $x(s)$ for the right-hand side of (6), we have

$$x(s) = \frac{1}{c^2}\int K^2(s,u)\,x(u)\,du.$$

Thus $x(s)$ is a characteristic function of K^2 with characteristic value $1/c^2$, and the result now follows from Lemma 4.

7·3. Characteristic systems. To avoid repetition, we shall assume throughout the rest of this chapter that K is a non-null Hermitian kernel. From now on, all our proofs will be independent of the Fredholm theory.

THEOREM 7·3·1. *Every characteristic value of K is real.*

Let $x = \lambda K x$, where x is non-null. Then

$$\lambda(Kx, x) = (x, x) \neq 0,$$

so that $\lambda(Kx, x)$ is real and non-zero. Since K is Hermitian,

$$(Kx, x) = (x, Kx) = \overline{(Kx, x)},$$

so that (Kx, x) is real. The characteristic value λ is therefore real.

THEOREM 7·3·2. *Characteristic functions of K belonging to distinct characteristic values are orthogonal to one another.*

Since $K^* = K$ and $\bar{\lambda} = \lambda$, this is a special case of Theorem 6·7·4.

THEOREM 7·3·3. *The characteristic values of K form a finite or enumerable sequence (λ_n), with no finite limit point. If we include each characteristic value in the sequence a number of times equal to its rank, then*

$$\sum_{n=1}^{\infty} \frac{1}{\lambda_n^2} \leqslant \| K \|^2 < \infty. \tag{1}$$

The first statement could be deduced from the fact that every characteristic value is a zero of the integral function $\delta(\lambda)$, the modified Fredholm determinant of K; the proof given below is independent of the Fredholm theory.

We have already seen (Corollary to Theorem 6·7·2) that there is a finite orthonormal base of characteristic functions for each characteristic value λ of K. If we choose such a base for each λ, and form their union, then, by Theorem 7·3·2, we obtain an orthonormal system, which we shall call a *full orthonormal system* of characteristic functions of the kernel K. If $x_1, x_2, \ldots, x_n$ are distinct members of such a system, belonging respectively to the characteristic values $\lambda_1, \lambda_2, \ldots, \lambda_n$ (not necessarily all different), then $(x_\nu \otimes x_\nu)$ $(1 \leqslant \nu \leqslant n)$ is an orthonormal system of functions of two variables. By Bessel's inequality (Theorem 4·2·1), we have

$$\begin{aligned} \| K \|^2 &\geqslant \sum_{\nu=1}^{n} | (K, x_\nu \otimes x_\nu) |^2 \\ &= \sum_{\nu=1}^{n} | (Kx_\nu, x_\nu) |^2 \\ &= \sum_{\nu=1}^{n} \frac{| (x_\nu, x_\nu) |^2}{\lambda_\nu^2} \\ &= \sum_{\nu=1}^{n} \frac{1}{\lambda_\nu^2}. \end{aligned}$$

Hence, if $| \lambda_\nu | \leqslant \alpha$ $(1 \leqslant \nu \leqslant n)$, we shall have

$$\| K \|^2 \geqslant \sum_{\nu=1}^{n} \frac{1}{\alpha^2} = \frac{n}{\alpha^2},$$

i.e. $n \leqslant \alpha^2 \| K \|^2$; we see thus that the system of characteristic values, repeated as required by the statement of the theorem, has only a finite number of members in any finite interval $(-\alpha, \alpha)$. The whole system is therefore at most enumerable, and so can be written as a sequence (λ_n); the inequality (1) follows at once.

A given orthonormal system (x_n) will be a full orthonormal system of characteristic functions of K if and only if (a) each x_n is a characteristic function of K, say with characteristic value λ_n, and (b) every $\mathfrak{L}^2$ characteristic function x of K is a finite linear combination of functions of the system (x_n). The necessity of the conditions is obvious. To prove their sufficiency, suppose that x is a non-null $\mathfrak{L}^2$ function such that

$$x = \lambda K x.$$

By condition (*b*), we can write

$$x = \alpha_1 x_1 + \ldots + \alpha_n x_n$$

for some n. Hence

$$x = \lambda K x = \lambda \sum_{\nu=1}^{n} \alpha_\nu K x_\nu,$$

i.e.

$$\sum_{\nu=1}^{n} \alpha_\nu x_\nu = \sum_{\nu=1}^{n} \frac{\lambda \alpha_\nu}{\lambda_\nu} x_\nu,$$

$$\sum_{\nu=1}^{n} \alpha_\nu \left(1 - \frac{\lambda}{\lambda_\nu}\right) x_\nu = 0.$$

Since the x_ν are orthonormal, they are linearly independent, whence

$$\alpha_\nu \left(1 - \frac{\lambda}{\lambda_\nu}\right) = 0 \quad (1 \leqslant \nu \leqslant n).$$

Since x is non-null, $\alpha_\nu \neq 0$ for some ν. For such values of ν, we have $\lambda_\nu = \lambda$; when $\lambda_\nu \neq \lambda$, we must have $\alpha_\nu = 0$. The characteristic value λ is therefore equal to one of the λ_ν, and we have

$$x = \sum_{\lambda_\nu = \lambda} \alpha_\nu x_\nu.$$

The system (x_n) consequently contains an orthonormal base for the characteristic functions belonging to each characteristic value of K, i.e. it is a full orthonormal system of characteristic functions of K.

Throughout the remainder of this chapter we shall suppose that the sequence (λ_n) is so indexed that

$$|\lambda_1| \leqslant |\lambda_2| \leqslant \ldots$$

and that each characteristic value is included in the sequence a number of times equal to its rank. A corresponding full orthonormal system of characteristic functions will be denoted by (x_n). By a *characteristic system* $(x_n; \lambda_n)$ of the kernel K we shall then understand any pair of sequences (x_n), (λ_n) satisfying these conditions.

The system (x_n) is not necessarily complete. It may even be finite, as we can see by taking $K = y \otimes y$, where $\| y \| = 1$; the equation $x = \lambda K x$ is equivalent to $x = \lambda (x, y)\, y$, so that x must be of the form αy for some constant α; substituting this in the equation, we get $\alpha y = \lambda\alpha y$, whence $\lambda = 1$. The single function y thus forms a full orthonormal system for this kernel.

7·4. Expansion theorems. We now come to a series of theorems concerning the expansion of the kernel, and of functions represented in a certain sense by the kernel, in terms of the characteristic values and characteristic functions. The first of these theorems is the analogue of the following result for finite matrices: if **K** is a Hermitian matrix, then there is a unitary matrix **U** such that $\mathbf{U}^{-1}\mathbf{K}\mathbf{U}$ is diagonal; in other words, by transforming to an orthonormal base of the vector space consisting of characteristic vectors of **K**, the matrix representing the operator **K** becomes diagonal.

THEOREM 7·4·1. *If $(x_n; \lambda_n)$ is a characteristic system of K, then*

$$K = {}^{\circ}\sum_{n=1}^{\infty} \frac{x_n \otimes x_n}{\lambda_n} \quad (\mathfrak{L}^2). \tag{1}$$

Since, by Theorem 7·3·3,

$$\sum_{n=1}^{\infty} \frac{1}{\lambda_n^2} < \infty,$$

and $(x_n \otimes x_n)$ is an orthonormal system, it follows from the Riesz–Fischer theorem (Theorem 4·3·2) for functions of two variables that the series on the right-hand side of (1) is convergent in $\mathfrak{L}^2$ to an $\mathfrak{L}^2$ kernel $Q(s,t)$, which clearly satisfies $Q(s,t) = {}^{\circ}\overline{Q(t,s)}$; by modifying $Q(s,t)$ on a null set, if necessary, we can ensure that it is Hermitian in the full sense. Write $R = K - Q$; our object is now to prove that $R(s,t) = {}^{\circ} 0$.

If y is an arbitrary $\mathfrak{L}^2$ function, it follows from Theorem 4·5·1 that

$$Qy = {}^{\circ}\sum_{n=1}^{\infty} \frac{(y, x_n)\, x_n}{\lambda_n} \quad (\mathfrak{L}^2). \tag{2}$$

In particular, if we take $y = x_m$ for some m, we obtain

$$Qx_m = {}^{\circ} \frac{x_m}{\lambda_m} \quad (m \geqslant 1),$$

whence $Rx_m = ^\circ 0$ for all m. If R were non-null, it would, by Theorem 7·2·1, possess a characteristic value λ_0 and a corresponding characteristic function x_0. We then have

$$x_0 = \lambda_0 R x_0,$$

whence

$$(x_0, x_m) = \lambda_0(Rx_0, x_m) = \lambda_0(x_0, Rx_m) = 0 \quad (m \geqslant 1). \tag{3}$$

Taking $y = x_0$ in (2), we obtain, using (3),

$$Qx_0 = ^\circ \sum_{n=1}^{\infty} \frac{(x_0, x_n)\, x_n}{\lambda_n} \quad (\mathfrak{L}^2)$$
$$= ^\circ 0.$$

Hence $$\lambda_0 K x_0 = \lambda_0 Q x_0 + \lambda_0 R x_0 = ^\circ 0 + x_0 = x_0,$$

i.e. x_0 is equivalent to a characteristic function of K; this contradicts (3), since we supposed that (x_n) was a full orthonormal system of characteristic functions of K. We must therefore have $R(s,t) = ^\circ 0$, and (1) is established.

One might expect that when the system (x_n) is finite, equivalence could be replaced by equality in (1). That this is not so can be seen from the following example: in $0 \leqslant s \leqslant 1$, $0 \leqslant t \leqslant 1$, let

$$K(s,t) = 1 \quad (s \neq \tfrac{1}{2},\ t \neq \tfrac{1}{2}),$$
$$= 2 \quad (s = \tfrac{1}{2} \text{ or } t = \tfrac{1}{2});$$

the only characteristic functions are then multiples of $x_1(s)$, where

$$x_1(s) = 1 \ \ (s \neq \tfrac{1}{2}), \quad x_1(s) = 2 \ \ (s = \tfrac{1}{2}),$$

the corresponding characteristic value being $\lambda_1 = 1$; however,

$$K(\tfrac{1}{2}, \tfrac{1}{2}) - x_1(\tfrac{1}{2})\, \overline{x_1(\tfrac{1}{2})} = -2 \neq 0.$$

COROLLARY. *Under the hypothesis of Theorem* 7·4·1,

$$\| K \|^2 = \sum_{n=1}^{\infty} \frac{1}{\lambda_n^2}.$$

This follows at once from (1) and Theorem 4·5·1.

If an $\mathfrak{L}^2$ function y can be expressed in the form $y = Kx$, where x is an $\mathfrak{L}^2$ function, y is sometimes said to be representable by K.

Another way of looking at the situation is to say that y belongs to the range of the operator defined by K. We shall show below that functions of this kind can be expressed as an infinite series in terms of a characteristic system of K.

THEOREM 7·4·2. *Let $(x_n; \lambda_n)$ be a characteristic system of K; if $y = Kx$, where x is an $\mathfrak{L}^2$ function, then*

$$y = {}^{\circ}\sum_{n=1}^{\infty} (y, x_n)\, x_n = \sum_{n=1}^{\infty} \frac{(x, x_n)}{\lambda_n} x_n, \tag{4}$$

the series being relatively uniformly absolutely convergent.

The two series in (4) are identical, since

$$(y, x_n) = (Kx, x_n) = (x, Kx_n) = \frac{(x, x_n)}{\lambda_n}.$$

The convergence of the series is proved as follows. By Cauchy's inequality,

$$\left\{ \sum_{\nu=n+1}^{m} \left| \frac{(x, x_\nu)}{\lambda_\nu} x_\nu(s) \right| \right\}^2 \leqslant \sum_{\nu=n+1}^{m} |(x, x_\nu)|^2 \sum_{\nu=n+1}^{m} \frac{|x_\nu(s)|^2}{\lambda_\nu^2}. \tag{5}$$

Since
$$\int K(s,t)\, x_n(t)\, dt = \frac{x_n(s)}{\lambda_n} \quad (n \geqslant 1),$$

Bessel's inequality (Theorem 4·2·1), applied to $K(s,t)$ as a function of t, gives

$$\sum_{n=1}^{\infty} \frac{|x_n(s)|^2}{\lambda_n^2} \leqslant \int |K(s,t)|^2\, dt = [k_1(s)]^2. \tag{6}$$

Hence, by (5),

$$\sum_{\nu=n+1}^{m} \left| \frac{(x, x_\nu)}{\lambda_\nu} x_\nu(s) \right| \leqslant k_1(s) \left\{ \sum_{\nu=n+1}^{m} |(x, x_\nu)|^2 \right\}^{\frac{1}{2}} \to 0 \quad (n, m \to \infty).$$

The series in (4) is therefore relatively uniformly absolutely convergent; it is consequently convergent in mean square to the same sum. By Theorems 4·5·1 and 7·4·1, this sum is $y(s)$; since, however, mean square limits are only unique up to equivalence, we cannot say more than that $y(s)$ is equivalent to the sum of the series.

To show that we cannot expect to replace equivalence by

equality in this theorem, we consider the following example. In $0 \leqslant s \leqslant 1$, $0 \leqslant t \leqslant 1$, take

$$K(s,t) = 1 \qquad (s \neq \tfrac{1}{2},\ t \neq \tfrac{1}{2}),$$
$$= 1 + t \quad (s = \tfrac{1}{2}),$$
$$= 1 + s \quad (t = \tfrac{1}{2}).$$

This is a real symmetric kernel, and is therefore Hermitian. It has the single characteristic value $\lambda_1 = 1$, the corresponding characteristic function being given by

$$x_1(s) = 1 \ \ (s \neq \tfrac{1}{2}), \quad x_1(s) = \tfrac{3}{2} \ \ (s = \tfrac{1}{2}).$$

Now take $x(t) = t$. If $y = Kx$, we have

$$y(s) = \tfrac{1}{2} \ \ (s \neq \tfrac{1}{2}), \quad y(s) = \tfrac{5}{6} \ \ (s = \tfrac{1}{2}).$$

On the other hand, the series (4), which reduces to the single term $(y, x_1)\, x_1(s)$, has the sum $\tfrac{1}{2}$ when $s \neq \tfrac{1}{2}$ and $\tfrac{3}{4}$ when $s = \tfrac{1}{2}$, so that there is a discrepancy at the point $s = \tfrac{1}{2}$.

We shall see in § 7·6 that equivalence can be replaced by equality when $K(s,t)$ is continuous. This can also be done if the full orthonormal system (x_n) is complete, for we can then argue as follows: we have

$$\left| y(s) - \sum_{\nu=1}^{n} \frac{(x, x_\nu)}{\lambda_\nu} x_\nu(s) \right| = \left| \int K(s,t) \left\{ x(t) - \sum_{\nu=1}^{n} (x, x_\nu)\, x_\nu(t) \right\} dt \right|$$
$$\leqslant k_1(s) \left\| x - \sum_{\nu=1}^{n} (x, x_\nu)\, x_\nu \right\|$$
$$\to 0 \quad (n \to \infty),$$

since for a complete orthonormal system we have

$$\sum_{n=1}^{\infty} (x, x_n)\, x_n = {}^{\circ} x \quad (\mathfrak{L}^2)$$

for an arbitrary $\mathfrak{L}^2$ function x (§ 4·3).

COROLLARY. *If x is an $\mathfrak{L}^2$ function such that $Kx = {}^{\circ}0$, then $(x, x_n) = 0$ $(n \geqslant 1)$, and conversely.*

We have $(Kx, x_n) = (x, x_n)/\lambda_n$, so that $Kx = {}^{\circ}0$ implies $(x, x_n) = 0$. The converse follows at once from Theorem 7·4·2.

THEOREM 7·4·3. *Let $(x_n; \lambda_n)$ be a characteristic system of K; if x and y are $\mathfrak{L}^2$ functions, then*

$$(Kx, y) = \sum_{n=1}^{\infty} \frac{(x, x_n)(x_n, y)}{\lambda_n}. \tag{7}$$

The result follows at once by forming the inner product of the equation

$$Kx = {}^\circ \sum_{n=1}^{\infty} \frac{(x, x_n)}{\lambda_n} x_n$$

with the function y; since the series is relatively uniformly convergent, we can perform this operation term by term.

Equation (7) is known as the *Hilbert formula*. By putting $y = x$ we obtain, as a particular case of the formula,

$$(Kx, x) = \sum_{n=1}^{\infty} \frac{|(x, x_n)|^2}{\lambda_n}. \tag{8}$$

This equation is the analogue of the algebraic theorem that a Hermitian form

$$\sum_{\mu, \nu=1}^{n} k_{\mu\nu} \bar{\xi}_\mu \xi_\nu$$

in a finite number of variables can be converted by a unitary transformation of the variables into the canonical form

$$\sum_{\nu=1}^{n} \kappa_\nu |\xi_\nu|^2,$$

the numbers κ_ν being the eigenvalues of the matrix $\mathbf{K} = [k_{\mu\nu}]$.

We now investigate the construction of a Hermitian $\mathfrak{L}^2$ kernel with given characteristic values and characteristic functions.

THEOREM 7·4·4. *Given a sequence (λ_n) of real numbers such that $|\lambda_n| \leqslant |\lambda_{n+1}|$ $(n \geqslant 1)$ and*

$$\sum_{n=1}^{\infty} \frac{1}{\lambda_n^2} < \infty, \tag{9}$$

and an orthonormal system (y_n), there exists a Hermitian $\mathfrak{L}^2$ kernel $K(s,t)$ having a characteristic system of the form $(x_n; \lambda_n)$, where $x_n = {}^\circ y_n$ $(n \geqslant 1)$. The kernel $K(s,t)$ is unique in the sense of equivalence.

Since $(y_n \otimes y_n)$ is an orthonormal system of functions of two variables, it follows as in the proof of Theorem 7·4·1 that the series

$$\sum_{n=1}^{\infty} \frac{y_n \otimes y_n}{\lambda_n}$$

is convergent in mean square to a Hermitian $\mathfrak{L}^2$ kernel $K(s,t)$. If x is an $\mathfrak{L}^2$ function, we have, by Theorem 4·5·1,

$$Kx = {}^\circ \sum_{n=1}^{\infty} \frac{(x, y_n)\, y_n}{\lambda_n} \quad (\mathfrak{L}^2);$$

in particular, by taking $x = y_m$, we obtain

$$\lambda_m K y_m = {}^\circ y_m \quad (m \geqslant 1).$$

Writing $x_m = \lambda_m K y_m$ $(m \geqslant 1)$, we see that each x_m is a characteristic function of K, and that the system (x_m) is orthonormal.

Now let x be an arbitrary characteristic function of K; then

$$x = \lambda K x = {}^\circ \sum_{n=1}^{\infty} \frac{\lambda (x, y_n)}{\lambda_n} y_n \quad (\mathfrak{L}^2);$$

whence
$$(x, y_m) = \frac{\lambda}{\lambda_m} (x, y_m) \quad (m \geqslant 1).$$

Thus we must have $(x, y_m) = 0$ unless $\lambda = \lambda_m$, so that

$$x = {}^\circ \sum_{\lambda_m = \lambda} (x, y_m)\, y_m = {}^\circ \sum_{\lambda_m = \lambda} (x, x_m)\, x_m.$$

Consequently

$$x = \lambda K x = \sum_{\lambda_m = \lambda} (x, x_m)\, \lambda_m K x_m = \sum_{\lambda_m = \lambda} (x, x_m)\, x_m;$$

thus x is equal to a finite linear combination of the x_m, which therefore form a full orthonormal system of characteristic functions of K, i.e. $(x_n; \lambda_n)$ is a characteristic system of K.

Finally, by Theorem 7·4·1,

$$K = {}^\circ \sum_{n=1}^{\infty} \frac{x_n \otimes x_n}{\lambda_n} \quad (\mathfrak{L}^2),$$

so that K is unique in the sense of equivalence.

Theorem 7·4·4 is best possible in the sense that we cannot replace equivalence by equality in the equations $x_n = {}^\circ y_n$; in other words, not every infinite orthonormal system can be a full orthonormal system of characteristic functions for a Hermitian $\mathfrak{L}^2$ kernel with a given infinite sequence of characteristic values. For, let $(x_n; \lambda_n)$ be a characteristic system of the Hermitian $\mathfrak{L}^2$ kernel $K(s,t)$; by (6), we have

$$\sum_{n=1}^{\infty} \frac{|x_n(s)|^2}{\lambda_n^2} \leqslant \int |K(s,t)|^2 dt < \infty$$

for all s. If we now put

$$y_n(s) = x_n(s) \quad (s \neq s_0), \quad y_n(s) = \lambda_n \quad (s = s_0),$$

then (y_n) is an orthonormal system, but the series

$$\sum_{n=1}^{\infty} \frac{|y_n(s)|^2}{\lambda_n^2}$$

is divergent when $s = s_0$; thus $(y_n; \lambda_n)$ is not a characteristic system for any Hermitian $\mathfrak{L}^2$ kernel.

7·5. The iterated kernels. For the iterates of a Hermitian $\mathfrak{L}^2$ kernel the expansion theorems of § 7·4 can be considerably improved. We begin by constructing characteristic systems for the iterated kernels.

THEOREM 7·5·1. *Let $(x_n; \lambda_n)$ be a characteristic system of the Hermitian $\mathfrak{L}^2$ kernel K, and let $p > 1$. Then $(x_n; \lambda_n^p)$ is a characteristic system of K^p.*

From the equation

$$x_n = \lambda_n K x_n \quad (n \geqslant 1),$$

it follows at once that

$$x_n = \lambda_n^p K^p x_n \quad (n \geqslant 1),$$

so that x_n is a characteristic function of K^p with characteristic value λ_n^p. It remains to be shown that every characteristic function of K^p can be expressed as a finite linear combination of the (x_n); to do this, we use an argument that generalizes the proof of Lemma 4 of § 7·2. Let $x = \lambda^p K^p x \neq 0$, and denote the pth roots of unity by $1, \omega, \omega^2, \ldots, \omega^{p-1}$. Define y_ν by the equations

$$py_\nu = x + \omega^\nu \lambda K x + \omega^{2\nu} \lambda^2 K^2 x + \ldots + \omega^{(p-1)\nu} \lambda^{p-1} K^{p-1} x \qquad (0 \leqslant \nu \leqslant p-1).$$

Adding these p equations, we obtain

$$x = y_0 + y_1 + \dots + y_{p-1}. \tag{1}$$

Also $$p(\omega^\nu \lambda K y_\nu) = \omega^\nu \lambda K x + \omega^{2\nu} \lambda^2 K^2 x + \dots + \omega^{p\nu} \lambda^p K^p x$$
$$= p y_\nu \quad (0 \leqslant \nu \leqslant p-1),$$

since $\omega^{p\nu} = 1$ and $\lambda^p K^p x = x$. Thus, for each ν, either $y_\nu = 0$ or y_ν is a characteristic function of K with characteristic value $\omega^\nu \lambda$. In either case y_ν is a finite linear combination of the (x_n), so that, by (1), x is a finite linear combination of the (x_n). This completes the proof.†

THEOREM 7·5·2. *The expansion*

$$K^2(s,t) = {}^\circ\sum_{n=1}^{\infty} \frac{x_n(s)\,\overline{x_n(t)}}{\lambda_n^2} \tag{2}$$

holds in the following sense: the series is relatively uniformly absolutely convergent in s for each value of t, and in t for each value of s; the sum of the series is equivalent to $K^2(s,t)$ as a function of t for each value of s, and as a function of s for each value of t. Finally the series is dominatedly convergent in (s,t).

The first statement follows from Theorem 7·4·2 by taking $x(u) = K(u,t)$ for fixed t, so that $y(s) = K^2(s,t)$, and by using the fact that $K^2(s,t)$ is Hermitian. To prove that the series is dominatedly convergent, we observe that, by Cauchy's inequality,

$$\sum_{\nu=1}^{n} \left| \frac{x_\nu(s)\,\overline{x_\nu(t)}}{\lambda_\nu^2} \right| \leqslant \left\{ \sum_{\nu=1}^{n} \frac{|x_\nu(s)|^2}{\lambda_\nu^2} \right\}^{\frac{1}{2}} \left\{ \sum_{\nu=1}^{n} \frac{|x_\nu(t)|^2}{\lambda_\nu^2} \right\}^{\frac{1}{2}}, \tag{3}$$

and, by § 7·4 (6), the right-hand side of (3) is dominated by $k_1(s)\,k_1(t)$, which is an $\mathfrak{L}^2$ kernel. It follows from the general theory of vector lattices (Birkhoff (1948), p. 243) that the series is almost relatively uniformly convergent in (s,t); whether it is relatively uniformly convergent in the strict sense is not known.

The example discussed after the proof of Theorem 7·4·2 shows that equivalence cannot in general be replaced by equality in (2).

† We note that this process of deriving characteristic functions of K from those of K^p does not depend on the Hermitian character of K. The same idea can be used to adapt the entire Fredholm theory to cases where $K(s,t)$ itself is not an $\mathfrak{L}^2$ kernel but has an iterate $K^p(s,t)$ that is one.

A simple calculation shows that in this case $K^2(\frac{1}{2},\frac{1}{2})=\frac{7}{3}$, whereas the right-hand side of (2) reduces to $x_1(\frac{1}{2})\overline{x_1(\frac{1}{2})}=\frac{9}{4}$. However, if the orthonormal system (x_n) is complete, it follows from the remark made after the proof of Theorem 7·4·2 that it is then possible to replace equivalence by equality.

THEOREM 7·5·3. *If $p>2$, then*

$$K^p(s,t)=\sum_{n=1}^{\infty}\frac{x_n(s)\overline{x_n(t)}}{\lambda_n^p}, \tag{4}$$

the series being relatively uniformly absolutely convergent in (s,t).

By Theorem 7·4·1,

$$\left\|K-\sum_{\nu=1}^{n}\frac{x_\nu\otimes x_\nu}{\lambda_\nu}\right\|\to 0 \quad (n\to\infty).$$

Hence we also have, for $r\geqslant 2$,

$$\begin{aligned}\left\|K^r-\sum_{\nu=1}^{n}\frac{x_\nu\otimes x_\nu}{\lambda_\nu^r}\right\|&=\left\|K^{r-1}\left(K-\sum_{\nu=1}^{n}\frac{x_\nu\otimes x_\nu}{\lambda_\nu}\right)\right\|\\&\leqslant\|K^{r-1}\|.\left\|K-\sum_{\nu=1}^{n}\frac{x_\nu\otimes x_\nu}{\lambda_\nu}\right\|\\&\to 0\quad (n\to\infty).\end{aligned}$$

Consequently, if $p>2$,

$$\begin{aligned}&\left|K^p(s,t)-\sum_{\nu=1}^{n}\frac{x_\nu(s)\overline{x_\nu(t)}}{\lambda_\nu^p}\right|\\&=\left|\iint K(s,u)\left\{K^{p-2}(u,v)-\sum_{\nu=1}^{n}\frac{x_\nu(u)\overline{x_\nu(v)}}{\lambda_\nu^{p-2}}\right\}K(v,t)\,du\,dv\right|\\&\leqslant k_1(s)\,k_2(t)\left\|K^{p-2}-\sum_{\nu=1}^{n}\frac{x_\nu\otimes x_\nu}{\lambda_\nu^{p-2}}\right\|\\&\to 0\quad (n\to\infty).\end{aligned}\tag{5}$$

The inequality in (5) follows from Theorem 2·1·3.

THEOREM 7·5·4. *If x is an $\mathfrak{L}^2$ function, and $y=K^px$, where $p>1$, then*

$$y(s)=\sum_{n=1}^{\infty}(y,x_n)\,x_n(s)=\sum_{n=1}^{\infty}\frac{(x,x_n)}{\lambda_n^p}x_n(s), \tag{6}$$

the series being relatively uniformly absolutely convergent.

The two series in (6) are clearly identical. Relatively uniform absolute convergence follows at once from Theorem 7·4·2, applied to the kernel K^p. To prove that the sum is equal to $y(s)$ for all s, we remark that

$$\begin{aligned}\left|y(s)-\sum_{\nu=1}^{n}\frac{(x,x_\nu)}{\lambda_\nu^p}x_\nu(s)\right| &= \left|\int\int K(s,u)\left\{K^{p-1}(u,v)-\sum_{\nu=1}^{n}\frac{x_\nu(u)\overline{x_\nu(v)}}{\lambda_\nu^{p-1}}\right\}x(v)\,du\,dv\right| \\ &\leqslant k_1(s)\,\|x\|\,.\left\|K^{p-1}-\sum_{\nu=1}^{n}\frac{x_\nu\otimes x_\nu}{\lambda_\nu^{p-1}}\right\| \\ &\to 0 \quad (n\to\infty). \end{aligned} \tag{7}$$

The inequality in (7) follows from $|(Lx,z)|\leqslant\|L\|\,.\,\|x\|\,.\,\|z\|$ by taking

$$L=K^{p-1}-\sum_{\nu=1}^{n}\frac{x_\nu\otimes x_\nu}{\lambda_\nu^{p-1}}$$

and $z(u)=\overline{K(s,u)}$.

7·6. Continuous kernels. The results of the last two sections can be improved somewhat when the kernel $K(s,t)$ is continuous. We begin with a lemma, usually known as Dini's theorem,† belonging to the theory of functions of a real variable.

LEMMA. *If $\{u_n(s)\}$ is a monotone sequence of real continuous functions, convergent in the closed interval $a\leqslant s\leqslant b$ to a continuous function $u(s)$, then the convergence of the sequence is uniform in the interval.*

We may suppose without loss of generality that $u(s)=0$ $(a\leqslant s\leqslant b)$ and that $u_n(s)\geqslant u_{n+1}(s)$ for all n and s. Choose $\epsilon>0$. Given s_0, there is an integer n_0 such that

$$0\leqslant u_{n_0}(s_0)<\tfrac{1}{2}\epsilon;$$

since $u_{n_0}(s)$ is continuous at s_0, we can choose δ so that

$$|u_{n_0}(s)-u_{n_0}(s_0)|<\tfrac{1}{2}\epsilon \quad (|s-s_0|<\delta).$$

Hence
$$0\leqslant u_{n_0}(s)<\epsilon \quad (|s-s_0|<\delta),$$

and *a fortiori*

$$0\leqslant u_n(s)<\epsilon \quad (|s-s_0|<\delta,\ n\geqslant n_0). \tag{1}$$

† Dini (1878), pp. 110–12.

Thus with each point s_0 of $a \leqslant s \leqslant b$ we can associate an open interval $I = (s_0 - \delta, s_0 + \delta)$ and an integer n_0 such that (1) holds. By the Borel covering theorem, we can cover the closed interval $a \leqslant s \leqslant b$ by a finite number of the intervals I, say $(I_\rho)_{1 \leqslant \rho \leqslant r}$. Let $(n_\rho)_{1 \leqslant \rho \leqslant r}$ be the associated integers, and write

$$N = \max(n_1, n_2, \ldots, n_r).$$

We then have

$$0 \leqslant u_n(s) < \epsilon \quad (a \leqslant s \leqslant b,\ n \geqslant N);$$

in other words, $u_n(s) \to 0$ uniformly in $a \leqslant s \leqslant b$.

THEOREM 7·6·1. *If $K(s,t)$ is continuous, and $p > 1$, then*

$$K^p(s,t) = \sum_{n=1}^{\infty} \frac{x_n(s)\,\overline{x_n(t)}}{\lambda_n^p}, \tag{2}$$

the series being uniformly absolutely convergent in (s,t).

We distinguish the cases $p > 2$ and $p = 2$. When $p > 2$, the result follows at once from § 7·5 (4), together with the continuity of the function $k_1(s)\,k_2(t)$. Since $K(s,t)$ is Hermitian, we have $k_2(t) = k_1(t)$, whence we see that the assumption that $k_1(s)$ is bounded would be sufficient to prove the result in this case.

When $p = 2$, it follows from the proof of Theorem 7·5·2 that the series is uniformly convergent in t for each value of s, and in s for each value of t; in fact, the boundedness of $k_1(s)$ is sufficient to establish this much. Since $K(s,t)$ is continuous, its characteristic functions $x_n(s)$ are all continuous; the sum of the series is therefore continuous in s and t separately. This is sufficient, together with the results of Theorem 7·5·2, to show that equation (2) holds for all (s,t) without exception. To show that the series is uniformly convergent in (s,t) we remark that putting $t = s$ in (2) gives

$$K^2(s,s) = \sum_{n=1}^{\infty} \frac{|x_n(s)|^2}{\lambda_n^2} \quad (a \leqslant s \leqslant b). \tag{3}$$

The partial sums of (3) form a monotone sequence of continuous functions converging to the continuous function $K^2(s,s)$; by the lemma, the series (3) is uniformly convergent. The uniform absolute convergence of (2) now follows from the inequality

$$\left\{ \sum_{\nu=n+1}^{m} \left| \frac{x_\nu(s)\,\overline{x_\nu(t)}}{\lambda_\nu^2} \right| \right\}^2 \leqslant \sum_{\nu=n+1}^{m} \frac{|x_\nu(s)|^2}{\lambda_\nu^2} \sum_{\nu=n+1}^{m} \frac{|x_\nu(t)|^2}{\lambda_\nu^2}.$$

It is clear that if $K(s,t)$ is continuous, and the series

$$\sum_{n=1}^{\infty} \frac{x_n(s)\,\overline{x_n(t)}}{\lambda_n} \tag{4}$$

is uniformly convergent, even in s and t separately, then its sum is $K(s,t)$ for all (s,t); if $K(s,t)$ is a general $\mathfrak{L}^2$ kernel, and (4) is convergent almost everywhere, its sum is equivalent to $K(s,t)$. However, even if $K(s,t)$ is continuous, the series (4) does not necessarily converge for all (s,t); an example can be constructed as follows.

Let $f(s)$ be an even real continuous function of period 2π, and let

$$K(s,t)=f(s-t) \quad (0 \leqslant s \leqslant 2\pi,\ 0 \leqslant t \leqslant 2\pi).$$

Suppose that $f(s)$ has the formal Fourier series

$$\tfrac{1}{2}a_0+\sum_{n=1}^{\infty} a_n \cos ns,$$

where $a_n \neq 0$ for all n. Then

$$\left(\frac{1}{\sqrt{(2\pi)}}, \frac{\cos s}{\sqrt{\pi}}, \frac{\sin s}{\sqrt{\pi}}, \frac{\cos 2s}{\sqrt{\pi}}, \frac{\sin 2s}{\sqrt{\pi}}, \ldots\right)$$

is a full orthonormal system of characteristic functions of K, the corresponding sequence of characteristic values being

$$\left(\frac{1}{\pi a_0}, \frac{1}{\pi a_1}, \frac{1}{\pi a_1}, \frac{1}{\pi a_2}, \frac{1}{\pi a_2}, \ldots\right).$$

We suppose furthermore that

$$|a_0| \geqslant |a_1| \geqslant |a_2| \geqslant \ldots,$$

so that the characteristic values are arranged in the customary order. We then have, formally,

$$\sum_{n=1}^{\infty} \frac{x_n(s)\,\overline{x_n(t)}}{\lambda_n} = \tfrac{1}{2}a_0+\sum_{n=1}^{\infty} a_n(\cos ns \cos nt+\sin ns \sin nt)$$

$$= \tfrac{1}{2}a_0+\sum_{n=1}^{\infty} a_n \cos n\,(s-t), \tag{5}$$

which is the Fourier series of $f(s-t)$. An example† given by Salem (1954) shows that there is a function $f(s)$ with all the properties described above and possessing a Fourier series that diverges for certain values of s; the series (4) is therefore divergent on certain lines of the form $s-t=\text{const.}$

THEOREM 7·6·2. *If $K(s,t)$ is a continuous Hermitian kernel, $x(s)$ is a continuous function, and $y=Kx$, then*

$$y(s)=\sum_{n=1}^{\infty}(y,x_n)\,x_n(s)=\sum_{n=1}^{\infty}\frac{(x,x_n)}{\lambda_n}\,x_n(s), \tag{6}$$

the series being uniformly absolutely convergent.

The uniform absolute convergence of the series follows at once from the proof of Theorem 7·4·2. Since the functions $x_n(s)$ are all continuous, the sum of the series is continuous; since, by Theorem 7·4·2, it is equivalent to the continuous function $y(s)$, (6) holds for all s without exception.

It is clear from the proof that the result holds under substantially weaker hypotheses; it would be sufficient to assume that $k_1(s)$ is continuous and that $x(s)$ is an $\mathfrak{L}^2$ function.

7·7. Definite kernels and Mercer's theorem. Let K be a Hermitian $\mathfrak{L}^2$ kernel. We shall say that K is *non-negative definite* if $(Kx,x)\geqslant 0$ for every $\mathfrak{L}^2$ function x, and that K is *positive definite* if in addition $(Kx,x)=0$ implies that x is null. *Non-positive definite* and *negative definite* kernels are defined similarly.

The following theorem is now almost trivial.

THEOREM 7·7·1. *A non-null kernel K is non-negative definite if and only if all its characteristic values are positive; K is positive definite if and only if the above condition is satisfied and, in addition, some (and therefore every) full orthonormal system of characteristic functions of K is complete.*

By Theorem 7·4·3,

$$(Kx,x)=\sum_{n=1}^{\infty}\frac{|(x,x_n)|^2}{\lambda_n}. \tag{1}$$

† The example suggested by Hellinger and Toeplitz (1927), pp. 1521–22, is not altogether satisfactory, since the condition $|a_0|\geqslant|a_1|\geqslant\ldots$ does not hold for it.

Hence, if all $\lambda_n > 0$, we must have $(Kx, x) \geqslant 0$ for all x; also, if any λ_n is negative, $(Kx_n, x_n) = \lambda_n^{-1} < 0$. This proves the first part of the theorem.

If K is non-negative definite, it follows from (1) that $(Kx, x) = 0$ if and only if $(x, x_n) = 0$ for all n; thus K will be positive definite if and only if the vanishing of (x, x_n) for all n implies that $x =^\circ 0$, in other words, if and only if the orthonormal system (x_n) is complete.

We shall now show that the expansion theorem (Theorem 7·4·1) can be greatly improved if K is both continuous and definite; this result was first proved by Mercer (1909).

THEOREM 7·7·2. *Let $K(s,t)$ be continuous, non-negative definite, and not identically zero. Then the series*

$$\sum_{n=1}^{\infty} \frac{1}{\lambda_n} \tag{2}$$

is convergent, and

$$K(s,t) = \sum_{n=1}^{\infty} \frac{x_n(s)\,\overline{x_n(t)}}{\lambda_n}, \tag{3}$$

the series being uniformly absolutely convergent in (s,t).

Since $K(s, t)$ is Hermitian, $K(s, s)$ is real for all s. We shall show first that $K(s, s) \geqslant 0$. Suppose that for some s_0 we have $K(s_0, s_0) = -\epsilon < 0$. Since $K(s,t)$ is continuous, its real part $\Re[K(s,t)]$ is also continuous and there is a neighbourhood $|s - s_0| < \delta$, $|t - t_0| < \delta$ of (s_0, t_0) in which

$$\Re[K(s,t)] < -\tfrac{1}{2}\epsilon.$$

We define an $\mathfrak{L}^2$ function $x(t)$ by taking $x(t) = 1$ in $(s_0 - \delta, s_0 + \delta)$ and $x(t) = 0$ outside this interval. Since $x(t)$ is real for all t, we have

$$\begin{aligned}(Kx, x) &= \iint \Re[K(s,t)]\, x(s)\, x(t)\, ds\, dt \\ &= \int_{s_0-\delta}^{s_0+\delta} ds \int_{s_0-\delta}^{s_0+\delta} \Re[K(s,t)]\, dt \\ &\leqslant -2\epsilon\delta^2 < 0,\end{aligned}$$

contradicting the hypothesis that K is non-negative definite. We must therefore have $K(s, s) \geqslant 0$ for all s.

For any fixed n for which λ_n is defined, let us write

$$R_n(s,t) = K(s,t) - \sum_{\nu=1}^{n} \frac{x_\nu(s)\overline{x_\nu(t)}}{\lambda_\nu}.$$

If x is a characteristic function of R_n with characteristic value λ, we have

$$x = \lambda R_n x = \lambda K x - \sum_{\nu=1}^{n} \frac{\lambda}{\lambda_\nu}(x, x_\nu)\, x_\nu. \tag{4}$$

Hence, if $1 \leqslant \mu \leqslant n$,

$$(x, x_\mu) = (\lambda K x, x_\mu) - \frac{\lambda}{\lambda_\mu}(x, x_\mu) = 0, \tag{5}$$

so that (4) reduces to $x = \lambda K x$. Thus x is a characteristic function of K, and therefore a finite linear combination of the (x_ν). Since, by (5), x is orthogonal to $x_1, x_2, \ldots, x_n$, it must be a finite linear combination of $x_{n+1}, x_{n+2}, \ldots$. Conversely, when $\mu > n$, x_μ (if it is defined at all) is a characteristic function of R_n with characteristic value λ_μ. Thus $(x_{n+1}, x_{n+2}, \ldots)$ is a full orthonormal system for R_n, the corresponding system of characteristic values being $(\lambda_{n+1}, \lambda_{n+2}, \ldots)$. Since these are all positive, $R_n(s,t)$ is non-negative definite, so that, by what has been proved above, $R_n(s,s) \geqslant 0$ for all s; in other words,

$$\sum_{\nu=1}^{n} \frac{|x_\nu(s)|^2}{\lambda_\nu} \leqslant K(s,s). \tag{6}$$

Integrating (6) with respect to s, we obtain

$$\sum_{\nu=1}^{n} \frac{1}{\lambda_\nu} \int |x_\nu(s)|^2\, ds = \sum_{\nu=1}^{n} \frac{1}{\lambda_\nu} \leqslant \int K(s,s)\, ds = \tau(K).$$

The series $\sum_{n=1}^{\infty} \lambda_n^{-1}$ is therefore convergent.

It also follows from (6) that the series

$$\sum_{n=1}^{\infty} \frac{|x_n(s)|^2}{\lambda_n}$$

is convergent for all s. By Cauchy's inequality,

$$\left\{ \sum_{\nu=n+1}^{m} \left| \frac{x_\nu(s)\overline{x_\nu(t)}}{\lambda_\nu} \right| \right\}^2 \leqslant \sum_{\nu=n+1}^{m} \frac{|x_\nu(s)|^2}{\lambda_\nu} \sum_{\nu=n+1}^{m} \frac{|x_\nu(t)|^2}{\lambda_\nu}. \tag{7}$$

Choose $\epsilon > 0$ and take a fixed value s_0 of s; there will then be an integer n_0 such that

$$\sum_{\nu=n+1}^{m} \frac{|x_\nu(s_0)|^2}{\lambda_\nu} < \epsilon \quad (n, m \geqslant n_0).$$

We also have
$$\sum_{\nu=n+1}^{m} \frac{|x_\nu(t)|^2}{\lambda_\nu} \leqslant K(t,t) \leqslant M,$$

where M is some positive constant. Hence, by (7),

$$\sum_{\nu=n+1}^{m} \left| \frac{x_\nu(s_0)\,\overline{x_\nu(t)}}{\lambda_\nu} \right| \leqslant (\epsilon M)^{\frac{1}{2}} \quad (n, m \geqslant n_0).$$

The series
$$\sum_{n=1}^{\infty} \frac{x_n(s)\,\overline{x_n(t)}}{\lambda_n} \tag{8}$$

is therefore uniformly absolutely convergent in t for each s, so that its sum $Q(s,t)$ is a continuous function of t for each s. A similar argument shows that $Q(s,t)$ is a continuous function of s for each t. Since, by Theorem 7·4·1, the series (8) is convergent in mean square to $K(s,t)$, $Q(s,t)$ must be equivalent to the continuous function $K(s,t)$. It follows without difficulty from the continuity properties that we have shown $Q(s,t)$ to possess that it is equal to $K(s,t)$ for all (s,t) without exception.

We still have to prove that (8) is uniformly absolutely convergent in (s,t). We now know that

$$\sum_{n=1}^{\infty} \frac{|x_n(s)|^2}{\lambda_n} = K(s,s). \tag{9}$$

The partial sums of the series on the left-hand side of (9) form a monotone sequence of continuous functions convergent to the continuous function $K(s,s)$. By Dini's theorem (§ 7·6, lemma) the series is uniformly convergent in s. The uniform absolute convergence of (8) now follows from the inequality (7).

7·8. Extremal properties of characteristic values. Let $K(s,t)$ be a non-null Hermitian kernel, and let $(x_n; \lambda_n)$ be a characteristic system for K. Let us write (λ_n^+) and (λ_n^-) for the subsequences of (λ_n) formed by the positive and negative characteristic values respectively; either of these may be empty, and either or both may be finite. We also write $\kappa_n = 1/\lambda_n$, $\kappa_n^+ = 1/\lambda_n^+$, $\kappa_n^- = 1/\lambda_n^-$. Finally, let us denote by (x_n^+) and (x_n^-) the corresponding subsequences of (x_n).

Our aim in this section is to show that the characteristic values of K possess certain extremal properties, which are useful both for the numerical calculation of the characteristic values and characteristic functions and for determining the behaviour of the characteristic values when the kernel K is subjected to small perturbations.†

In this section we shall use the notation $\{f(x): P(x)\}$ to denote the set of all the values of the function $f(x)$ for those values of x that satisfy the condition $P(x)$. Thus $\sup\{f(x): P(x)\}$ will mean the same as $\sup_{P(x)} f(x)$. The notation is more convenient than the usual one when the condition $P(x)$ is complicated. We use the notation 'inf' for the greatest lower bound.

THEOREM 7·8·1. *If the sequence* (κ_n^+) *is not empty, then*

$$\kappa_1^+ = \sup\{(Kx, x) : \| x \| = 1\}. \tag{1}$$

If κ_2^+ *exists, then*

$$\kappa_2^+ = \sup\{(Kx, x) : \| x \| = 1, (x, x_1^+) = 0\}. \tag{2}$$

Generally, if κ_n^+ *exists, then*

$$\kappa_n^+ = \sup\{(Kx, x) : \| x \| = 1, (x, x_1^+) = (x, x_2^+) = \dots = (x, x_{n-1}^+) = 0\}.$$

Similarly, if the sequence (κ_n^-) *is not empty, then*

$$\kappa_1^- = \inf\{(Kx, x) : \| x \| = 1\},$$

$$\kappa_2^- = \inf\{(Kx, x) : \| x \| = 1, (x, x_1^-) = 0\},$$

and so on, as long as the sequence continues.

Finally,

$$|\kappa_1| = \sup\{|(Kx, x)| : \| x \| = 1\} = \sup\{\| Kx \| : \| x \| = 1\};$$

if κ_2 *exists, then*

$$|\kappa_2| = \sup\{|(Kx, x)| : \| x \| = 1, (x, x_1) = 0\}$$

$$= \sup\{\| Kx \| : \| x \| = 1, (x, x_1) = 0\},$$

and so on.

† Bückner (1952), Courant & Hilbert (1953), ch. 6.

Write $\gamma_1 = \sup\{(Kx, x) : \| x \| = 1\}.$

If $x = x_1^+$, then $(Kx, x) = \kappa_1^+ (x, x) = \kappa_1^+,$

whence we must have $\gamma_1 \geqslant \kappa_1^+$. On the other hand, if x is an arbitrary $\mathfrak{L}^2$ function of norm 1, we have, by the Hilbert formula (Theorem 7·4·3),

$$(Kx, x) = \sum_{n=1}^{\infty} \kappa_n \mid (x, x_n) \mid^2$$

$$\leqslant \sum_{n=1}^{\infty} \kappa_n^+ \mid (x, x_n^+) \mid^2$$

$$\leqslant \kappa_1^+ \sum_{n=1}^{\infty} \mid (x, x_n^+) \mid^2$$

$$\leqslant \kappa_1^+ \| x \|^2 = \kappa_1^+;$$

hence we must also have $\gamma_1 \leqslant \kappa_1^+$, and (1) follows.

Now let $$\gamma_2 = \sup\{(Kx, x) : \| x \| = 1, (x, x_1^+) = 0\}. \tag{3}$$

If $x = x_2^+$, which satisfies both the conditions laid down for x in (3), then

$$(Kx, x) = \kappa_2^+ (x, x) = \kappa_2^+,$$

whence $\gamma_2 \geqslant \kappa_2^+$. On the other hand, if x is an arbitrary $\mathfrak{L}^2$ function of norm 1 and orthogonal to x_1^+, the Hilbert formula gives us

$$(Kx, x) \leqslant \sum_{n=2}^{\infty} \kappa_n^+ \mid (x, x_n^+) \mid^2 \leqslant \kappa_2^+ \sum_{n=2}^{\infty} \mid (x, x_n^+) \mid^2 \leqslant \kappa_2^+,$$

whence $\gamma_2 \leqslant \kappa_2^+$, and (2) follows.

The remaining statements of the theorem can clearly be proved by arguments of the same kind.

The results of Theorem 7·8·1 are analogous to certain extremal properties of the principal axes of an n-dimensional central quadric. Thus (1) corresponds to the fact that the smallest principal axis of an ellipsoid cuts the surface in points whose distance from the centre is a minimum, and (2) to the fact that the points nearest to the centre on the intersection of the ellipsoid and a plane through the centre perpendicular to the smallest principal axis lie on the middle principal axis. The lengths of the semi-axes in question correspond to $\sqrt{(\lambda_1^+)}$ and $\sqrt{(\lambda_2^+)}$.

The expressions for the characteristic values given by Theorem 7·8·1 have certain disadvantages; for instance, the formula for λ_n^+ involves the characteristic functions $(x_1^+, x_2^+, \ldots, x_{n-1}^+)$, which must therefore be known before the formula can be used. For certain purposes† it is desirable to have formulae that do not involve the earlier characteristic functions; these are provided by the following theorem, which exhibits the characteristic values as solutions of certain 'minimax' problems. This theorem, like Theorem 7·8·1, is analogous to certain results concerning the principal axes of central quadrics; the formulation of these is left to the reader.

THEOREM 7·8·2. *Suppose that* $\lambda_1^+, \lambda_2^+, \ldots, \lambda_k^+$ *all exist. Let* $p_1, p_2, \ldots, p_{k-1}$ *be* $\mathfrak{L}^2$ *functions, and write*

$$\mu^+(K; p_1, p_2, \ldots, p_{k-1}) = \sup\{(Kx, x) : \| x \| = 1, (x, p_1) = \ldots = (x, p_{k-1}) = 0\}.$$

Then
$$\kappa_k^+ = \inf \mu^+(K; p_1, p_2, \ldots, p_{k-1}), \tag{4}$$
the lower bound in (4) *being taken over all possible choices of* $p_1, p_2, \ldots, p_{k-1}$.

Similarly, if $\lambda_1^-, \lambda_2^-, \ldots, \lambda_k^-$ *all exist, and*

$$\mu^-(K; p_1, p_2, \ldots, p_{k-1}) = \inf\{(Kx, x) : \| x \| = 1, (x, p_1) = \ldots = (x, p_{k-1}) = 0\},$$

then
$$\kappa_k^- = \sup \mu^-(K; p_1, p_2, \ldots, p_{k-1}), \tag{5}$$
the upper bound in (5) *being taken over all possible choices of the* $\mathfrak{L}^2$ *functions* $p_1, p_2, \ldots, p_{k-1}$.

Given $p_1, p_2, \ldots, p_{k-1}$, we can form a linear combination

$$x = \alpha_1 x_1^+ + \ldots + \alpha_k x_k^+$$

of $x_1^+, \ldots, x_k^+$ such that

$$\| x \| = 1, \quad (x, p_1) = \ldots = (x, p_{k-1}) = 0.$$

We then have

$$(Kx, x) = \sum_{\nu=1}^{k} \kappa_\nu^+ \, | (x, x_\nu^+) |^2 = \sum_{\nu=1}^{k} | \alpha_\nu |^2 \kappa_\nu^+ \geqslant \kappa_k^+ \sum_{\nu=1}^{k} | \alpha_\nu |^2 = \kappa_k^+,$$

† Courant & Hilbert (1953) pp. 132–4.

whence $$\mu^+(K; p_1, p_2, \ldots, p_{k-1}) \geqslant \kappa_k^+.$$

Since κ_k^+ is independent of $p_1, p_2, \ldots, p_{k-1}$, it follows that

$$\inf \mu^+(K; p_1, p_2, \ldots, p_{k-1}) \geqslant \kappa_k^+. \tag{6}$$

On the other hand, if we take $p_\nu = x_\nu^+$ $(1 \leqslant \nu \leqslant k-1)$, and x is an arbitrary $\mathfrak{L}^2$ function such that

$$\| x \| = 1, \quad (x, p_1) = \ldots = (x, p_{k-1}) = 0,$$

then $$(Kx, x) \leqslant \sum_{n=k}^{\infty} \kappa_n^+ \mid (x, x_n^+) \mid^2 \leqslant \kappa_k^+ \sum_{n=k}^{\infty} \mid (x, x_n^+) \mid^2 \leqslant \kappa_k^+,$$

whence $$\mu^+(K; x_1^+, x_2^+, \ldots, x_{k-1}^+) \leqslant \kappa_k^+,$$

so that $$\inf \mu^+(K; p_1, p_2, \ldots, p_{k-1}) \leqslant \kappa_k^+. \tag{7}$$

The first result now follows from (6) and (7).

The result for the negative characteristic values is proved similarly.

There is, in general, no result of this kind for the sequence $(\mid \kappa_n \mid)$; suppose, for instance, that K has exactly two characteristic values, and that $\kappa_2 = -\kappa_1$. If we take $p_1 = x_1 + x_2$, then the $\mathfrak{L}^2$ functions of unit norm orthogonal to p_1 are those of the form

$$\alpha(x_1 - x_2) + y,$$

where $(y, x_1) = (y, x_2) = 0$, and $2 \mid \alpha \mid^2 + \| y \|^2 = 1$. If x is any such function, then

$$(Kx, x) = \kappa_1 \mid (x, x_1) \mid^2 + \kappa_2 \mid (x, x_2) \mid^2 = (\kappa_1 + \kappa_2) \mid \alpha \mid^2 = 0,$$

so that

$$\mu(K; p_1) = \sup \{ \mid (Kx, x) \mid : \| x \| = 1, (x, p_1) = 0 \} = 0,$$

and, since $\mu \geqslant 0$ for all p_1,

$$\inf \mu(K; p_1) = 0,$$

which has no relevance to the value of $\mid \kappa_2 \mid$.

The following theorem of Ky Fan (1949) will be used in Chapter VIII.

THEOREM 7·8·3. *Suppose that $\lambda_1^+, \lambda_2^+, \ldots, \lambda_k^+$ all exist. Then*

$$\sum_{\nu=1}^{k} \kappa_\nu^+ = \sup \sum_{\nu=1}^{k} (Ky_\nu, y_\nu),$$

the upper bound being taken over all orthonormal systems $(y_1, y_2, \ldots, y_k)$ *containing k elements.*

By the Hilbert formula (Theorem 7·4·3),

$$(Ky_\mu, y_\mu) = \sum_{\nu=1}^{\infty} \kappa_\nu \,|\, (y_\mu, x_\nu) \,|^2 \leqslant \sum_{\nu=1}^{\infty} \kappa_\nu^+ \,|\, (y_\mu, x_\nu^+) \,|^2, \tag{8}$$

where y_μ is any member of a given orthonormal system $(y_1, y_2, \ldots, y_k)$. We write (8) in the form

$$(Ky_\mu, y_\mu) \leqslant \kappa_k^+ \sum_{\nu=1}^{\infty} |\, (y_\mu, x_\nu^+) \,|^2 + \sum_{\nu=1}^{k} (\kappa_\nu^+ - \kappa_k^+) \,|\, (y_\mu, x_\nu^+) \,|^2$$
$$+ \sum_{\nu=k+1}^{\infty} (\kappa_\nu^+ - \kappa_k^+) \,|\, (y_\mu, x_\nu^+) \,|^2,$$

whence, since $\| y_\mu \| = 1$ and $\kappa_\nu^+ \leqslant \kappa_k^+$ $(\nu > k)$,

$$(Ky_\mu, y_\mu) \leqslant \kappa_k^+ + \sum_{\nu=1}^{k} (\kappa_\nu^+ - \kappa_k^+) \,|\, (y_\mu, x_\nu^+) \,|^2 \quad (1 \leqslant \mu \leqslant k).$$

Summing over μ, we have, after some rearrangement,

$$\sum_{\mu=1}^{k} \kappa_\mu^+ - \sum_{\mu=1}^{k} (Ky_\mu, y_\mu) \geqslant \sum_{\mu=1}^{k} (\kappa_\mu^+ - \kappa_k^+) - \sum_{\mu=1}^{k} \sum_{\nu=1}^{k} (\kappa_\nu^+ - \kappa_k^+) \,|\, (y_\mu, x_\nu^+) \,|^2$$
$$\geqslant \sum_{\mu=1}^{k} (\kappa_\mu^+ - \kappa_k^+) - \sum_{\nu=1}^{k} (\kappa_\nu^+ - \kappa_k^+)$$
$$= 0, \tag{9}$$

since
$$\sum_{\mu=1}^{k} |\, (y_\mu, x_\nu^+) \,|^2 \leqslant \| x_\nu^+ \|^2 = 1.$$

Since equality holds in (9) when we take $y_\mu = x_\mu^+$ $(1 \leqslant \mu \leqslant k)$, the result is established.

A similar result holds for the negative characteristic values.

7·9. The non-homogeneous equation. We shall now show how, by using the expansion theorems, it is possible to solve non-homogeneous integral equations with Hermitian kernels without appealing to the general Fredholm theory.

THEOREM 7·9·1. *Let* $(x_n; \lambda_n)$ *be a characteristic system of the non-null Hermitian kernel* K. *If* y *is an arbitrary* $\mathfrak{L}^2$ *function, and* λ *is not a characteristic value of* K, *the equation*

$$x = y + \lambda Kx \tag{1}$$

has a unique solution x, which satisfies the equation

$$x =^{\circ} y + \lambda \sum_{n=1}^{\infty} \frac{(y, x_n)}{\lambda_n - \lambda} x_n, \tag{2}$$

the series being relatively uniformly absolutely convergent. Denoting the right-hand side of (2) by x_0, we have

$$x = y + \lambda K x_0 = y + \lambda K y + \lambda^2 \sum_{n=1}^{\infty} \frac{(y, x_n)}{\lambda_n(\lambda_n - \lambda)} x_n, \tag{3}$$

the series again being relatively uniformly absolutely convergent.

Let x be an $\mathfrak{L}^2$ solution of (1). By Theorem 7·4·2, we have

$$Kx =^{\circ} \sum_{n=1}^{\infty} \frac{(x, x_n)}{\lambda_n} x_n,$$

the series being relatively uniformly absolutely convergent. Equation (1) thus implies that

$$x =^{\circ} y + \lambda \sum_{n=1}^{\infty} \frac{(x, x_n)}{\lambda_n} x_n. \tag{4}$$

Forming the inner product of both sides of (4) with x_m, we obtain

$$(x, x_m) = (y, x_m) + \frac{\lambda}{\lambda_m} (x, x_m) \quad (m \geqslant 1),$$

whence

$$(x, x_m) = \frac{\lambda_m}{\lambda_m - \lambda} (y, x_m) \quad (m \geqslant 1), \tag{5}$$

and, by (4),

$$x =^{\circ} y + \lambda \sum_{n=1}^{\infty} \frac{(y, x_n)}{\lambda_n - \lambda} x_n.$$

Thus any $\mathfrak{L}^2$ solution of (1) satisfies (2).

We shall now show, conversely, that if x_0 is defined by the right-hand side of (2), it is equivalent to a solution of (1). The series defining x_0 may be written

$$y + \lambda \sum_{n=1}^{\infty} \frac{\lambda_n}{\lambda_n - \lambda} \frac{(y, x_n)}{\lambda_n} x_n. \tag{6}$$

By Theorem 7·4·2, the series

$$\sum_{n=1}^{\infty} \frac{(y, x_n)}{\lambda_n} x_n$$

is relatively uniformly absolutely convergent. Since

$$\frac{\lambda_n}{\lambda_n-\lambda}\to 1 \quad (n\to\infty),$$

the series (6) is also relatively uniformly absolutely convergent. We may therefore write

$$x_0=y+\lambda\sum_{n=1}^{\infty}\frac{(y,x_n)}{\lambda_n-\lambda}x_n.$$

We then have

$$\begin{aligned}x_0-\lambda Kx_0&=y-\lambda Ky+\lambda\sum_{n=1}^{\infty}\frac{(y,x_n)}{\lambda_n-\lambda}x_n-\lambda^2\sum_{n=1}^{\infty}\frac{(y,x_n)}{\lambda_n(\lambda_n-\lambda)}x_n\\&=y-\lambda Ky+\lambda\sum_{n=1}^{\infty}\frac{(y,x_n)}{\lambda_n-\lambda}\left(1-\frac{\lambda}{\lambda_n}\right)x_n\\&=y-\lambda Ky+\lambda\sum_{n=1}^{\infty}\frac{(y,x_n)}{\lambda_n}x_n\\&=^{\circ}y,\end{aligned}$$

by Theorem 7·4·2 applied to Ky. Hence

$$x_0=^{\circ}y+\lambda Kx_0.$$

If we now write

$$x=y+\lambda Kx_0=y+\lambda Ky+\lambda^2\sum_{n=1}^{\infty}\frac{(y,x_n)}{\lambda_n(\lambda_n-\lambda)}x_n,$$

we have $x=^{\circ}x_0$, so that $\lambda Kx_0=\lambda Kx$ and

$$x=y+\lambda Kx;$$

thus x satisfies (1). This completes the proof.

Equivalence can be replaced by equality in (2) if either (*a*) $K(s,t)$ is continuous (or, more generally, $k_1(s)$ is continuous) or (*b*) the orthonormal system (x_n) is complete.

We can also express the resolvent H_λ of K as a series in terms of characteristic functions of K.

THEOREM 7·9·2. *If λ is not a characteristic value of the non-null Hermitian $\mathfrak{L}^2$ kernel K, then the resolvent H_λ of K exists, and*

$$H_\lambda=^{\circ}\sum_{n=1}^{\infty}\frac{x_n\otimes x_n}{\lambda_n-\lambda}\quad(\mathfrak{L}^2),\tag{7}$$

where $(x_n; \lambda_n)$ *is any characteristic system of* K. *Denoting the right-hand side of* (7) *by* H^0_λ, *we have*

$$H_\lambda = K + \lambda K^2 + \lambda^2 K H^0_\lambda K$$

$$= K + \lambda K^2 + \lambda^2 \sum_{n=1}^{\infty} \frac{x_n \otimes x_n}{\lambda_n^2(\lambda_n - \lambda)}, \tag{8}$$

the series being relatively uniformly absolutely convergent.

The series on the right-hand side of (7) is convergent in mean square, since $(x_n \otimes x_n)$ is an orthonormal system of functions of two variables and

$$\sum_{n=1}^{\infty} \frac{1}{|\lambda_n - \lambda|^2}$$

is convergent by comparison with the series $\Sigma(1/\lambda_n^2)$; it therefore defines an $\mathfrak{L}^2$ kernel $H^0_\lambda(s,t)$. We then have, by Theorem 7·4·1,

$$H^0_\lambda - K = {}^\circ\sum_{n=1}^{\infty} \left(\frac{1}{\lambda_n - \lambda} - \frac{1}{\lambda_n}\right)(x_n \otimes x_n) \quad (\mathfrak{L}^2)$$

$$= {}^\circ\lambda \sum_{n=1}^{\infty} \frac{x_n \otimes x_n}{\lambda_n(\lambda_n - \lambda)} \quad (\mathfrak{L}^2).$$

Let us write $\quad A_n = \sum_{\nu=1}^{n} \frac{x_\nu \otimes x_\nu}{\lambda_\nu - \lambda}, \quad B_n = \sum_{\nu=1}^{n} \frac{x_\nu \otimes x_\nu}{\lambda_\nu},$

so that $A_n \to H^0_\lambda$ $(\mathfrak{L}^2)$, $B_n \to K$ $(\mathfrak{L}^2)$. We then have

$$\| H^0_\lambda K - A_n B_n \|$$

$$= \| H^0_\lambda (K - B_n) + (H^0_\lambda - A_n) K - (H^0_\lambda - A_n)(K - B_n) \|$$

$$\leqslant \| H^0_\lambda \| . \| K - B_n \| + \| H^0_\lambda - A_n \| . \| K \| + \| H^0_\lambda - A_n \| . \| K - B_n \|$$

$$\to 0 \quad (n \to \infty),$$

i.e. $A_n B_n \to H^0_\lambda K$ $(\mathfrak{L}^2)$. Also

$$A_n B_n = \left(\sum_{\mu=1}^{n} \frac{x_\mu \otimes x_\mu}{\lambda_\mu - \lambda}\right)\left(\sum_{\nu=1}^{n} \frac{x_\nu \otimes x_\nu}{\lambda_\nu}\right)$$

$$= \sum_{\nu=1}^{n} \frac{x_\nu \otimes x_\nu}{\lambda_\nu(\lambda_\nu - \lambda)};$$

we therefore have

$$H^0_\lambda K = {}^\circ\sum_{n=1}^{\infty} \frac{x_n \otimes x_n}{\lambda_n(\lambda_n - \lambda)} \quad (\mathfrak{L}^2).$$

Consequently $$\lambda H_\lambda^0 K = ^\circ H_\lambda^0 - K; \tag{9}$$

similarly $$\lambda K H_\lambda^0 = ^\circ H_\lambda^0 - K. \tag{10}$$

Thus H_λ^0 satisfies the resolvent equation in the sense of equivalence. To rid ourselves of the equivalence signs, we put

$$H_\lambda = K + \lambda K^2 + \lambda^2 K H_\lambda^0 K. \tag{11}$$

By (9) and (10),

$$H_\lambda^0 = ^\circ K + \lambda K^2 + \lambda^2 K H_\lambda^0 K.$$

Thus $H_\lambda = ^\circ H_\lambda^0$, and

$$\begin{aligned} \lambda H_\lambda K &= \lambda K^2 + \lambda^2 K^3 + \lambda^3 K H_\lambda^0 K^2 \\ &= \lambda K^2 + \lambda^2 K (K + \lambda H_\lambda^0 K) K \\ &= \lambda K^2 + \lambda^2 K H_\lambda^0 K \\ &= H_\lambda - K. \end{aligned}$$

Similarly $\lambda K H_\lambda = H_\lambda - K$, so that H_λ is the resolvent of K. Finally, we obtain the series (8) for H_λ by substituting the expression (7) for H_λ^0 in (11); its relatively uniform absolute convergence follows by an argument of the type used in the proof of Theorem 7·5·3.

If $K(s,t)$ is continuous or the orthonormal system (x_n) is complete, we can replace the expression (8) by the simpler one

$$H_\lambda = K + \lambda \sum_{n=1}^{\infty} \frac{x_n \otimes x_n}{\lambda_n(\lambda_n - \lambda)}. \tag{12}$$

In the first case, the series will be uniformly absolutely convergent, and in the second case it will be convergent in the sense described in the enunciation of Theorem 7·5·2.

The expression (8) for the resolvent clearly exhibits its meromorphic character, and shows also that when we are dealing with a Hermitian kernel all the poles of the resolvent are simple.

Improved expansion theorems could be obtained for the resolvents of the iterated kernels, but we refrain from entering into the details.

We now consider the non-homogeneous equation in the case where λ is a characteristic value of the kernel.

THEOREM 7·9·3. *If λ is a characteristic value of the non-null Hermitian $\mathfrak{L}^2$ kernel K, $(u_1, u_2, \ldots, u_r)$ is a full set of characteristic functions belonging to λ, and y is an arbitrary $\mathfrak{L}^2$ function, the equation*

$$x = y + \lambda K x \tag{13}$$

has an $\mathfrak{L}^2$ solution x if and only if

$$(y, u_\rho) = 0 \quad (1 \leqslant \rho \leqslant r). \tag{14}$$

When (14) *holds and x is an $\mathfrak{L}^2$ solution of* (13), *then, for some constants $\alpha_1, \alpha_2, \ldots, \alpha_r$,*

$$x =^\circ y + \sum_{\rho=1}^{r} \alpha_\rho u_\rho + \lambda \sum_{\lambda_n \neq \lambda} \frac{(y, x_n)}{\lambda_n - \lambda} x_n, \tag{15}$$

where $(x_n; \lambda_n)$ is any characteristic system of K, the series being relatively uniformly absolutely convergent.

Conversely, if (14) *holds and $\alpha_1, \alpha_2, \ldots, \alpha_r$ are arbitrarily chosen constants, and if the right-hand side of* (15) *is denoted by x_0, then*

$$x = y + \lambda K x_0$$

is an $\mathfrak{L}^2$ solution of (13).

If x satisfies (13) we have, as in the proof of Theorem 7·9·1,

$$x =^\circ y + \lambda \sum_{n=1}^{\infty} \frac{(x, x_n)}{\lambda_n} x_n,$$

whence, for any fixed m,

$$(x, x_m) = (y, x_m) + \frac{\lambda}{\lambda_m} (x, x_m),$$

so that

$$(\lambda_m - \lambda)(x, x_m) = \lambda_m (y, x_m). \tag{16}$$

When $\lambda_m = \lambda$, (16) gives at once $(y, x_m) = 0$, so that (14) is a necessary condition for (13) to have an $\mathfrak{L}^2$ solution. When $\lambda_m \neq \lambda$ (16) gives, as in the proof of Theorem 7·9·1,

$$(x, x_m) = \frac{\lambda_m}{\lambda_m - \lambda} (y, x_m).$$

We therefore have

$$x =^\circ y + \sum_{\lambda_n = \lambda} \alpha_n x_n + \lambda \sum_{\lambda_n \neq \lambda} \frac{(y, x_n)}{\lambda_n - \lambda} x_n, \tag{17}$$

the coefficients α_n being undetermined.

Conversely, suppose that y satisfies (14). Choosing $\alpha_1, \alpha_2, \ldots, \alpha_r$ arbitrarily, define x_0 as the expression on the right-hand side of (17); the convergence of the series follows as in the proof of Theorem 7·9·1. We then have

$$\begin{aligned} x_0 - \lambda K x_0 &= y - \lambda K y + \lambda \sum_{\lambda_n \neq \lambda} \frac{(y, x_n)}{\lambda_n - \lambda}\left(1 - \frac{\lambda}{\lambda_n}\right) x_n \\ &= y - \lambda K y + \lambda \sum_{\lambda_n \neq \lambda} \frac{(y, x_n)}{\lambda_n} x_n \\ &= y - \lambda K y + \lambda \sum_{n=1}^{\infty} \frac{(y, x_n)}{\lambda_n} x_n \\ &= {}^{\circ} y, \end{aligned}$$

by Theorem 7·4·2 applied to Ky. Putting $x = y + \lambda K x_0$, and reasoning as in the proof of Theorem 7·9·1, we see that x is a solution of (13).

As in earlier results in this section, equivalence can be replaced by equality in (15) when $K(s, t)$ is continuous or the orthonormal system (x_n) is complete.

The particular solution x obtained by taking all the α_ρ to be 0 in (15) may be characterized as that making $\| x - y \|$ a minimum; for we clearly have

$$\| x - y \|^2 = | \lambda |^2 \sum_{\lambda_n \neq \lambda} \frac{| (y, x_n) |^2}{| \lambda_n - \lambda |^2} + \sum_{\lambda_n = \lambda} | \alpha_n |^2.$$

CHAPTER VIII

SINGULAR FUNCTIONS AND SINGULAR VALUES

8·1. Introductory remarks. We have seen in Chapter VII that a satisfactory theory can be constructed for linear integral equations of the second kind with Hermitian $\mathfrak{L}^2$ kernels. In particular, we have shown that such a kernel is determined, up to equivalence, by its system of characteristic values and characteristic functions. This result cannot be expected to hold for general $\mathfrak{L}^2$ kernels, for we have seen in § 7·1 that a non-null $\mathfrak{L}^2$ kernel may have no characteristic values at all.

The possibility remains, however, that there may exist some system of complex numbers and corresponding functions related to the kernel and determining it in a manner similar to that in which a Hermitian kernel is determined by its characteristic values and characteristic functions. This is in fact so, and in the present chapter we shall define and discuss the properties of the singular values and singular functions of a general $\mathfrak{L}^2$ kernel. The theory of these was first given by E. Schmidt (1907*a*) and was extended to general real kernels by Smithies (1937). Though the singular values and singular functions are not as closely related to the problem of the solution of integral equations as the characteristic values and characteristic functions, they have important applications; among these we shall consider the approximation of $\mathfrak{L}^2$ kernels by kernels of finite rank, the theory of normal kernels, and the theory of linear integral equations of the first kind.

8·2. Definitions and elementary properties. Let $K(s, t)$ be a non-null $\mathfrak{L}^2$ kernel, and let $K^*(s,t) = \overline{K(t,s)}$ be its adjoint kernel. If the non-null $\mathfrak{L}^2$ functions u and v and the complex number μ satisfy the relations

$$u = \mu K v, \quad v = \mu K^* u, \tag{1}$$

we shall call μ a *singular value* of K and $[u, v]$ a *pair of singular functions*† of K belonging to the singular value μ.

We shall base our discussion of singular values and singular functions on the theory of Hermitian kernels (Chapter VII); it is also possible to treat them directly.‡

THEOREM 8·2·1. *Let $[u, v]$ be a pair of singular functions belonging to the singular value μ of the non-null $\mathfrak{L}^2$ kernel K. Then μ is real, u is a characteristic function of KK^* belonging to the characteristic value μ^2, and v is a characteristic function of K^*K belonging to the same characteristic value.*

We have

$$u = \mu K v = \mu K(\mu K^* u) = \mu^2 K K^* u,$$

$$v = \mu K^* u = \mu K^*(\mu K v) = \mu^2 K^* K v,$$

so that u and v are characteristic functions of KK^* and K^*K respectively with the same characteristic value μ^2. This implies incidentally that KK^* and K^*K are non-null. Since

$$(K^*K)^* = K^*K^{**} = K^*K,$$

K^*K is Hermitian; similarly KK^* is Hermitian. If x is an arbitrary $\mathfrak{L}^2$ function,

$$(K^*Kx, x) = (Kx, Kx) = \| Kx \|^2 \geqslant 0,$$

so that K^*K is non-negative definite; similarly for KK^*. By Theorem 7·7·1, $\mu^2 > 0$, whence μ is real.

We remark that if $[u, v]$ is a pair of singular functions of K belonging to μ, then $[u, -v]$ is a pair belonging to $-\mu$. When, later on, we construct a full system of pairs of singular functions, we shall select only one of these two pairs, usually the one with the positive singular value. We shall therefore assume from now on that all singular values are taken positive, except when an explicit statement is made to the contrary.

THEOREM 8·2·2. *Let K be a non-null $\mathfrak{L}^2$ kernel; suppose that $\mu > 0$ and that u is a characteristic function of KK^* belonging to*

† We use square brackets to avoid confusion with the inner product (u, v) of u and v.

‡ Vergerio (1919), Mollerup (1924), Lewis (1950).

the characteristic value μ^2. *Then* $v=\mu K^*u$ *is a characteristic function of* K^*K *belonging to the same characteristic value, and* $[u, v]$ *is a pair of singular functions of* K *belonging to the singular value* μ.

A similar result holds when we are given that $v=\mu^2K^*Kv$, *where* v *is a non-null* $\mathfrak{L}^2$ *function.*

Since
$$u=\mu^2KK^*u,$$
we have
$$\mu K^*u=\mu^3K^*KK^*u=\mu^2K^*K(\mu K^*u),$$
i.e.
$$v=\mu^2K^*Kv.$$
Also
$$\mu Kv=\mu K(\mu K^*u)=\mu^2KK^*u=u,$$
so that v is non-null, and the result follows. The second part of the theorem is proved similarly.

THEOREM 8·2·3. *Let* K *be a non-null* $\mathfrak{L}^2$ *kernel, and let* $(u_n;\, \mu_n^2)$ *be a characteristic system of* KK^*, *where* $\mu_n>0$ $(n\geqslant 1)$. *If*
$$v_n=\mu_nK^*u_n \quad (n\geqslant 1),$$
then $(v_n;\, \mu_n^2)$ *is a characteristic system of* K^*K.

A similar result holds if we are given that $(v_n;\, \mu_n^2)$ *is a characteristic system of* K^*K.

We have, for all m and n,
$$\begin{aligned}(v_m, v_n)&=(\mu_mK^*u_m, \mu_nK^*u_n)\\ &=\mu_m\mu_n(KK^*u_m, u_n)\\ &=\frac{\mu_n}{\mu_m}(u_m, u_n)\\ &\left.\begin{aligned}&=1 \quad (m=n)\\ &=0 \quad (m\neq n).\end{aligned}\right\}\end{aligned}$$
The system (v_n) is therefore orthonormal. Now let v be any characteristic function of K^*K, with characteristic value μ^2, and write $u=\mu Kv$, where $\mu>0$. Then, by Theorem 8·2·2, u is a characteristic function of KK^*; it can therefore be expressed as a finite linear combination of the functions of the full orthonormal system (u_n), say
$$u=\sum_{\nu=1}^{n}\alpha_\nu u_\nu.$$
Then
$$v=\mu K^*u=\sum_{\nu=1}^{n}\mu\alpha_\nu K^*u_\nu=\sum_{\nu=1}^{n}\frac{\mu\alpha_\nu}{\mu_\nu}v_\nu,$$

so that v is a finite linear combination of the functions v_n. The system (v_n) is therefore (§ 7·3) a full orthonormal system for K^*K. The second part of the theorem is proved similarly.

If $(u_n;\mu_n^2)$ is a characteristic system of KK^* and $(v_n;\mu_n^2)$ is a characteristic system of K^*K, where $\mu_n>0$ $(n\geqslant 1)$ and the two systems are related by the equations

$$u_n=\mu_n Kv_n,\quad v_n=\mu_n K^*u_n \quad (n\geqslant 1),$$

we shall say that $[u_n, v_n]$ is a *full system* of pairs of singular functions of K or, more briefly, that $(u_n, v_n;\mu_n)$ is a *singular system* of K. We shall suppose the sequence (μ_n) of singular values of K to be arranged in non-decreasing order, so that

$$0<\mu_1\leqslant\mu_2\leqslant\ldots\leqslant\mu_n\leqslant\ldots.$$

8·3. Expansion theorems. Let $K(s,t)$ be a non-null $\mathfrak{L}^2$ kernel, and let $(u_n, v_n;\mu_n)$ be a singular system of K. In the present section we shall prove a series of expansion theorems similar to those of § 7·4.

THEOREM 8·3·1. *A necessary and sufficient condition for an $\mathfrak{L}^2$ function x to satisfy the condition $Kx=^\circ 0$ is that $(x,v_n)=0$ for all n; similarly, $K^*x=^\circ 0$ if and only if $(x,u_n)=0$ for all n.*

Since (v_n) is a full system of characteristic functions of K^*K, the condition that $(x,v_n)=0$ for all n is, by Theorem 7·4·2, Corollary, equivalent to $K^*Kx=^\circ 0$. This is clearly implied by $Kx=^\circ 0$; conversely, $K^*Kx=^\circ 0$ implies that

$$0=(K^*Kx,x)=(Kx,Kx)=\| Kx\|^2,$$

whence $Kx=^\circ 0$. The second part of the theorem is proved similarly.

THEOREM 8·3·2. *If x is an $\mathfrak{L}^2$ function and $y=Kx$, then*

$$y=^\circ\sum_{n=1}^{\infty}(y,u_n)\,u_n=\sum_{n=1}^{\infty}\frac{(x,v_n)}{\mu_n}u_n, \tag{1}$$

the series being relatively uniformly absolutely convergent. Similarly, if $z=K^*x$, then

$$z=^\circ\sum_{n=1}^{\infty}(z,v_n)\,v_n=\sum_{n=1}^{\infty}\frac{(x,u_n)}{\mu_n}v_n, \tag{2}$$

the series being relatively uniformly absolutely convergent.

Since

$$(y, u_n) = (Kx, u_n) = (x, K^*u_n) = \frac{(x, v_n)}{\mu_n} \quad (n \geqslant 1),$$

the two series in (1) are identical. We begin by showing that this series converges to $y = Kx$ in mean square. We have

$$\left\| y - \sum_{\nu=1}^{n} \frac{(x, v_\nu)}{\mu_\nu} u_\nu \right\|^2 = \left\| Kx - \sum_{\nu=1}^{n} \frac{(x, v_\nu)}{\mu_\nu} u_\nu \right\|^2$$

$$= \| Kx \|^2 - \sum_{\nu=1}^{n} \frac{\overline{(x, v_\nu)}\,(Kx, u_\nu)}{\mu_\nu}$$

$$- \sum_{\nu=1}^{n} \frac{(x, v_\nu)\,\overline{(Kx, u_\nu)}}{\mu_\nu}$$

$$+ \sum_{\nu, \rho=1}^{n} \frac{(x, v_\nu)\,\overline{(x, v_\rho)}\,(u_\nu, u_\rho)}{\mu_\nu \mu_\rho}$$

$$= \| Kx \|^2 - \sum_{\nu=1}^{n} \frac{(x, v_\nu)\,\overline{(x, v_\nu)}}{\mu_\nu^2}$$

$$= (K^*Kx, x) - \sum_{\nu=1}^{n} \frac{| (x, v_\nu) |^2}{\mu_\nu^2}$$

$$\to 0 \quad (n \to \infty),$$

by the Hilbert formula (Theorem 7·4·3) applied to K^*K. Next, we have

$$\int K(s, t)\, v_n(t)\, dt = \frac{u_n(s)}{\mu_n} \quad (n \geqslant 1),$$

so that, by Bessel's inequality, applied to $K(s, t)$ as a function of t for fixed s,

$$\sum_{\nu=1}^{n} \frac{| u_\nu(s) |^2}{\mu_\nu^2} \leqslant \int | K(s, t) |^2\, dt = [k_1(s)]^2. \tag{3}$$

Consequently, by Schwarz's inequality,

$$\left\{ \sum_{\nu=n+1}^{m} \frac{| (x, v_\nu) |}{\mu_\nu} | u_\nu(s) | \right\}^2 \leqslant \sum_{\nu=n+1}^{m} | (x, v_\nu) |^2 \sum_{\nu=n+1}^{m} \frac{| u_\nu(s) |^2}{\mu_\nu^2}$$

$$\leqslant [k_1(s)]^2 \sum_{\nu=n+1}^{m} | (x, v_\nu) |^2. \tag{4}$$

Since $\Sigma|(x, v_n)|^2$ is convergent, it follows from (4) that the series in (1) is relatively uniformly absolutely convergent; since it is also convergent in mean square to $y = Kx$, its sum is equivalent to Kx.

The second part of the theorem is proved similarly.

As in the case of Theorem 7·4·2, we can replace equivalence by equality in (1) if (v_n) is a complete orthonormal system, and we can do the same in (2) if (u_n) is complete. For

$$\left|y(s) - \sum_{\nu=1}^{n} \frac{(x, v_\nu)}{\mu_\nu} u_\nu(s)\right|^2 = \left|\int K(s,t)\left\{x(t) - \sum_{\nu=1}^{n}(x, v_\nu)\, v_\nu(t)\right\} dt\right|^2$$

$$\leqslant [k_1(s)]^2 \left\| x - \sum_{\nu=1}^{n} (x, v_\nu)\, v_\nu \right\|^2$$

$$\to 0 \quad (n \to \infty),$$

if (v_n) is complete.

COROLLARY 1. *If $K(s,t)$ is a continuous kernel, equivalence can be replaced by equality in* (1) *and* (2), *and the series are uniformly convergent.*

Uniform convergence in (1) follows at once from (4); since the functions $u_n(s)$ are continuous, the sum of the series is continuous, and it is equivalent to the continuous function $y(s)$. Hence the result for (1); the proof for (2) is similar.

COROLLARY 2. *If x and y are $\mathfrak{L}^2$ functions, then*

$$(Kx, y) = \sum_{n=1}^{\infty} \frac{(x, v_n)(u_n, y)}{\mu_n}. \tag{5}$$

This follows at once by taking the inner product with y of the equation

$$Kx = {}^\circ\sum_{n=1}^{\infty} \frac{(x, v_n)}{\mu_n} u_n.$$

Equation (5) is the analogue of the Hilbert formula.

Our next result is the analogue of Theorem 7·4·1 (the bilinear expansion).

THEOREM 8·3·3. *If $(u_n, v_n; \mu_n)$ is a singular system of the non-null $\mathfrak{L}^2$ kernel $K(s,t)$, then*

$$K = {}^\circ\sum_{n=1}^{\infty} \frac{u_n \otimes v_n}{\mu_n} \quad (\mathfrak{L}^2). \tag{6}$$

By the inequality (3) in the proof of Theorem 8·3·2,

$$\sum_{\nu=1}^{n} \frac{|u_\nu(s)|^2}{\mu_\nu^2} \leqslant \int |K(s,t)|^2 dt;$$

integrating with respect to s, we obtain

$$\sum_{\nu=1}^{n} \frac{1}{\mu_\nu^2} \leqslant \|K\|^2,$$

which implies that the series $\Sigma\mu_n^{-2}$ is convergent. Let us now write

$$K_n = \sum_{\nu=1}^{n} \frac{u_\nu \otimes v_\nu}{\mu_\nu}. \tag{7}$$

If $n < m$, we have

$$\|K_n - K_m\|^2 = \left\| \sum_{\nu=n+1}^{m} \frac{u_\nu \otimes v_\nu}{\mu_\nu} \right\|^2 = \sum_{\nu=n+1}^{m} \frac{1}{\mu_\nu^2} \to 0 \quad (n, m \to \infty).$$

The sequence (K_n) is therefore convergent in mean square to an $\mathfrak{L}^2$ kernel K_0. If now x is an arbitrary $\mathfrak{L}^2$ function, we have, by Theorem 8·3·2,

$$\|K_n x - Kx\| = \left\| \sum_{\nu=1}^{n} \frac{(x, v_\nu)}{\mu_\nu} u_\nu - Kx \right\| \to 0 \quad (n \to \infty);$$

on the other hand,

$$\|K_n x - K_0 x\| \leqslant \|K_n - K_0\| . \|x\| \to 0 \quad (n \to \infty).$$

Hence $K_0 x = {}^\circ Kx$ for all x, whence it follows, by the argument used in the proof of Theorem 7·2·1, Lemma 1, that $K_0 = {}^\circ K$; this gives the required results.

Corollary. *Under the assumptions of Theorem* 8·3·3,

$$\sum_{n=1}^{\infty} \frac{1}{\mu_n^2} = \|K\|^2. \tag{8}$$

This follows at once from Theorem 4·5·1.

8·4. The approximation theorem. In §3·3 we outlined a proof that every $\mathfrak{L}^2$ kernel can be approximated in mean square as closely as we please by kernels of finite rank. This result is also an immediate consequence of Theorem 8·3·3, since the kernels K_n defined by § 8·3 (7) are all of finite rank. We shall now show that, for given n, the best mean square approximation to an $\mathfrak{L}^2$ kernel K by kernels of rank not exceeding n is given by the kernel K_n.

THEOREM 8·4·1. *Let $(u_n, v_n; \mu_n)$ be a singular system for the $\mathfrak{L}^2$ kernel K, and write*

$$K_n = \sum_{\nu=1}^{n} \frac{u_\nu \otimes v_\nu}{\mu_\nu}.$$

If $a_1, \dots, a_n, b_1, \dots, b_n$ are arbitrary $\mathfrak{L}^2$ functions, then

$$\left\| K - \sum_{\nu=1}^{n} a_\nu \otimes b_\nu \right\|^2 \geqslant \| K - K_n \|^2 = \sum_{\nu=n+1}^{\infty} \frac{1}{\mu_\nu^2}. \tag{1}$$

By the Corollary to Theorem 8·3·3, it is sufficient to prove that

$$\left\| K - \sum_{\nu=1}^{n} a_\nu \otimes b_\nu \right\|^2 \geqslant \| K \|^2 - \sum_{\nu=1}^{n} \frac{1}{\mu_\nu^2}. \tag{2}$$

We may suppose that the set $(b_1, b_2, \dots, b_n)$ is orthonormal; for, if it is not, we form a linearly equivalent orthonormal set, and express the b_ν in terms of it. The expression

$$\sum_{\nu=1}^{n} a_\nu \otimes b_\nu$$

is then equivalent to a similar expression with not more than n terms in which the functions playing the role of the b_ν are orthonormal. We thus have

$$\begin{aligned}
&\left\| K - \sum_{\nu=1}^{n} a_\nu \otimes b_\nu \right\|^2 \\
&= \left(K - \sum_{\mu=1}^{n} a_\mu \otimes b_\mu,\ K - \sum_{\nu=1}^{n} a_\nu \otimes b_\nu \right) \\
&= \| K \|^2 - \sum_{\mu=1}^{n} (a_\mu, K b_\mu) - \sum_{\nu=1}^{n} (K b_\nu, a_\nu) + \sum_{\mu,\nu=1}^{n} (a_\mu, a_\nu)(b_\nu, b_\mu) \\
&= \| K \|^2 - \sum_{\nu=1}^{n} (a_\nu, K b_\nu) - \sum_{\nu=1}^{n} (K b_\nu, a_\nu) + \sum_{\nu=1}^{n} (a_\nu, a_\nu) \\
&= \| K \|^2 + \sum_{\nu=1}^{n} (a_\nu - K b_\nu, a_\nu - K b_\nu) - \sum_{\nu=1}^{n} (K b_\nu, K b_\nu) \\
&\geqslant \| K \|^2 - \sum_{\nu=1}^{n} \| K b_\nu \|^2.
\end{aligned} \tag{3}$$

By (2) and (3), it is sufficient to prove that

$$\sum_{\nu=1}^{n} \| Kb_\nu \|^2 \leqslant \sum_{\nu=1}^{n} \frac{1}{\mu_\nu^2}.$$

To prove this, we need only remark that

$$\sum_{\nu=1}^{n} \| Kb_\nu \|^2 = \sum_{\nu=1}^{n} (K^*Kb_\nu, b_\nu) \leqslant \sum_{\nu=1}^{n} \frac{1}{\mu_\nu^2},$$

by Theorem 7·8·3, applied to the Hermitian kernel K^*K.

8·5. Hermitian kernels. We shall now indicate how some of the results of the present chapter reduce to those of Chapter VII in the case where the kernel is Hermitian.

Let $(x_n; \lambda_n)$ be a characteristic system for the Hermitian $\mathfrak{L}^2$ kernel K. Since $$x_n = \lambda_n K x_n = \lambda_n K^* x_n,$$
$[x_n, x_n]$ is a pair of singular functions of K belonging to the singular value λ_n; we have here suspended the convention that singular values are to be taken positive. By Theorem 7·5·1, $(x_n; \lambda_n^2)$ is a characteristic system for the kernel

$$K^2 = KK^* = K^*K.$$

By Theorem 8·2·3, modified to allow negative singular values, $(x_n, x_n; \lambda_n)$ is a singular system for K, and the results of § 8·3 then reduce to results already proved in § 7·4.

Alternatively, we can maintain the convention of taking the singular values positive by using $(x_n, x_n \operatorname{sgn} \lambda_n; |\lambda_n|)$ as our singular system for K, where $\operatorname{sgn} \lambda_n = \pm 1$ according as λ_n is positive or negative.

8·6. Normal kernels. The class of normal kernels, which we discuss in the present section, includes the class of Hermitian kernels, and possesses most of its properties. In fact normality is, roughly speaking, a necessary and sufficient condition for an $\mathfrak{L}^2$ kernel to possess a bilinear expansion in terms of a characteristic system (Theorems 8·6·5 and 8·6·8).

In § 2·2 we defined a normal kernel as an $\mathfrak{L}^2$ kernel K for which $KK^* = K^*K$, i.e.

$$\int K(s, u) \overline{K(t, u)}\, du = \int \overline{K(u, s)}\, K(u, t)\, du$$

for all (s, t), and we remarked that every Hermitian $\mathfrak{L}^2$ kernel is normal. We shall also say that an $\mathfrak{L}^2$ kernel K is *almost normal* if $KK^* =^\circ K^*K$; we shall see later (Theorem 8·6·7) that every almost normal kernel is equivalent to a normal kernel.

THEOREM 8·6·1. *A necessary and sufficient condition for an $\mathfrak{L}^2$ kernel K to be almost normal is that*

$$\| Kx \| = \| K^*x \|$$

for every $\mathfrak{L}^2$ function x.

If K is almost normal and x is an $\mathfrak{L}^2$ function, we have

$$\| Kx \|^2 = (Kx, Kx) = (K^*Kx, x) = (KK^*x, x)$$

$$= (K^*x, K^*x) = \| K^*x \|^2,$$

whence $\| Kx \| = \| K^*x \|$. Conversely, if $\| Kx \| = \| K^*x \|$ for every $\mathfrak{L}^2$ function x, the above argument shows that

$$(K^*Kx, x) = (KK^*x, x),$$

i.e. the Hermitian kernel $H = K^*K - KK^*$ satisfies $(Hx, x) = 0$ for all x. If H were non-null, it would have a non-null characteristic function x with characteristic value λ, say, and we should have

$$0 = (Hx, x) = \lambda^{-1}(x, x),$$

a contradiction; hence H is null, i.e. $K^*K =^\circ KK^*$.

THEOREM 8·6·2. *An arbitrary $\mathfrak{L}^2$ kernel K can be expressed uniquely in the form*

$$K = G + iH,$$

where G and H are Hermitian $\mathfrak{L}^2$ kernels. A necessary and sufficient condition for K to be normal is that $GH = HG$.

Given K, we put

$$G = \tfrac{1}{2}(K + K^*), \quad H = \frac{1}{2i}(K - K^*);$$

G and H are clearly Hermitian, and we have $K = G + iH$. On the other hand, if $K = G_0 + iH_0$, where G_0 and H_0 are Hermitian, then $K^* = G_0 - iH_0$, whence

$$G_0 = \tfrac{1}{2}(K + K^*) = G, \quad H_0 = \frac{1}{2i}(K - K^*) = H.$$

Finally, we have

$$KK^* = (G+iH)(G-iH) = G^2+H^2-i(GH-HG),$$
$$K^*K = (G-iH)(G+iH) = G^2+H^2+i(GH-HG);$$

the last part of the theorem follows at once.

We may regard G and H as being respectively the real and imaginary parts of K; thus a kernel is normal if and only if its real and imaginary parts (in this sense) commute.

THEOREM 8·6·3. *Let K be a normal $\mathfrak{L}^2$ kernel. Then* (i) *if x is an $\mathfrak{L}^2$ function such that $x=\lambda Kx$, then $x=\bar{\lambda}K^*x$, and* (ii) *if x_1 and x_2 are $\mathfrak{L}^2$ characteristic functions of K with distinct characteristic values λ_1 and λ_2, then $(x_1, x_2)=0$.*

To prove (i), we remark that

$$\begin{aligned}\| x-\bar{\lambda}K^*x \|^2 &= (x-\bar{\lambda}K^*x, x-\bar{\lambda}K^*x)\\ &= (x,x)-\bar{\lambda}(K^*x,x)-\lambda(x,K^*x)+\lambda\bar{\lambda}(K^*x,K^*x)\\ &= (x,x)-\bar{\lambda}(x,Kx)-\lambda(Kx,x)+\lambda\bar{\lambda}(KK^*x,x)\\ &= (x,x)-\lambda(Kx,x)-\bar{\lambda}(x,Kx)+\lambda\bar{\lambda}(K^*Kx,x)\\ &= (x,x)-\lambda(Kx,x)-\bar{\lambda}(x,Kx)+\lambda\bar{\lambda}(Kx,Kx)\\ &= \| x-\lambda Kx \|^2\\ &= 0,\end{aligned}$$

whence

$$x \overset{\circ}{=} \bar{\lambda}K^*x. \tag{1}$$

To convert the equivalence into an equality, we note that (1) implies

$$\lambda Kx = \lambda\bar{\lambda}KK^*x;$$

since $\lambda Kx = x$, it follows that

$$x = \lambda\bar{\lambda}KK^*x = \lambda\bar{\lambda}K^*Kx = \bar{\lambda}K^*(\lambda Kx) = \bar{\lambda}K^*x.$$

Part (ii) of the theorem now follows at once from Theorem 6·7·4.

With Theorem 8·6·3 at our disposal, we can define full orthonormal systems of characteristic functions and characteristic systems for normal kernels in exactly the same way as we did in § 7·3 for Hermitian kernels. However, we do not yet know whether a non-null normal kernel has any characteristic values at all. Instead of proving this existence theorem directly, we shall appeal to the theory of singular functions, thus indirectly using the theory of Hermitian kernels. Our fundamental theorem is as follows.

THEOREM 8·6·4. *Let K be a non-null normal $\mathfrak{L}^2$ kernel, and let $(x_n; \lambda_n)$ be a characteristic system of K. Then $(x_n, x_n \operatorname{sgn} \lambda_n; |\lambda_n|)$ is a singular system of K.*

The notation $\operatorname{sgn} \lambda$ used in the statement of the theorem is defined by the equations

$$\operatorname{sgn} \lambda = \frac{\lambda}{|\lambda|} \quad (\lambda \neq 0), \quad \operatorname{sgn} 0 = 0.$$

We split the proof, which is fairly long, into a number of stages.

(i) Let μ be a singular value of K, and let $[u_1, v_1], [u_2, v_2], \ldots, [u_r, v_r]$ be those members of some full system of pairs of singular functions that belong to the singular value μ. Then we have, for $1 \leqslant \rho \leqslant r$,

$$KK^*(\mu K u_\rho) = \mu K(K^* K u_\rho) = \mu K(KK^* u_\rho)$$
$$= \mu K(\mu^{-2} u_\rho) = \mu^{-2}(\mu K u_\rho);$$

also, by Theorem 8·6·1,

$$\| \mu K u_\rho \| = |\mu| . \| K u_\rho \| = |\mu| . \| K^* u_\rho \| = \| \mu K^* u_\rho \| = \| v_\rho \| = 1,$$

so that $\mu K u_\rho$ is non-null. Thus $\mu K u_\rho$ is a characteristic function of KK^* with characteristic value μ^2; it is therefore a linear combination of $u_1, u_2, \ldots, u_r$, say

$$\mu K u_\rho = \sum_{\tau=1}^{r} w_{\tau\rho} u_\tau \quad (1 \leqslant \rho \leqslant r). \tag{2}$$

We also have

$$(\mu K u_\rho, \mu K u_\sigma) = \mu^2 (K^* K u_\rho, u_\sigma) = \mu^2 (KK^* u_\rho, u_\sigma) = (u_\rho, u_\sigma) = \delta_{\rho\sigma}. \tag{3}$$

Substituting in (3) from (2), we obtain

$$\delta_{\rho\sigma} = \sum_{\tau=1}^{r} \sum_{\nu=1}^{r} (w_{\tau\rho} u_\tau, w_{\nu\sigma} u_\nu)$$
$$= \sum_{\tau=1}^{r} \sum_{\nu=1}^{r} w_{\tau\rho} \overline{w}_{\nu\sigma} (u_\tau, u_\nu)$$
$$= \sum_{\tau=1}^{r} w_{\tau\rho} \overline{w}_{\tau\sigma}. \tag{4}$$

Denoting the $r \times r$ matrix $[w_{\tau\rho}]$ by $\mathbf{W}$, we can write (4) in the form $\mathbf{W}^*\mathbf{W} = \mathbf{I}$; in other words, $\mathbf{W}$ is a unitary matrix and we have $\mathbf{W}^{-1} = \mathbf{W}^*$.

(ii) We now require two lemmas on unitary matrices. We give their proofs for completeness.

LEMMA 1. *The eigenvalues of a unitary matrix* $\mathbf{W}$ *are of absolute value* 1.

Let $\mathbf{W}\mathbf{x} = \alpha\mathbf{x}$, where $\mathbf{x}$ is a non-zero vector. Then

$$\mathbf{x}^*\mathbf{x} = \mathbf{x}^*\mathbf{I}\mathbf{x} = \mathbf{x}^*\mathbf{W}^*\mathbf{W}\mathbf{x} = (\mathbf{W}\mathbf{x})^*(\mathbf{W}\mathbf{x}) = (\alpha\mathbf{x})^*(\alpha\mathbf{x}) = \bar{\alpha}\alpha\mathbf{x}^*\mathbf{x};$$

since $\mathbf{x}^*\mathbf{x} \neq 0$, we have $\bar{\alpha}\alpha = 1$, $|\alpha| = 1$.

LEMMA 2. *Given a unitary matrix* $\mathbf{W}$, *there is a unitary matrix* $\mathbf{T}$ *such that* $\mathbf{D} = \mathbf{T}^*\mathbf{W}\mathbf{T} = \mathbf{T}^{-1}\mathbf{W}\mathbf{T}$ *is a diagonal matrix. The diagonal elements of* $\mathbf{D}$ *are the eigenvalues of* $\mathbf{W}$.

We proceed inductively; supposing the result already known for $(r-1) \times (r-1)$ matrices, we shall prove it for $r \times r$ matrices. Since the result holds trivially when $r = 1$, the induction can be started.

Let $\mathbf{s}_1$ be an eigenvector of $\mathbf{W}$ with eigenvalue α_1, and let $(\mathbf{s}_1, \mathbf{s}_2, \ldots, \mathbf{s}_r)$ be an orthonormal base (with $\mathbf{s}_1$ as first element) of the complex vector space of r dimensions; we then have $\mathbf{s}_\rho^*\mathbf{s}_\sigma = \delta_{\rho\sigma}$. The matrix $\mathbf{S} = [\mathbf{s}_1, \mathbf{s}_2, \ldots, \mathbf{s}_r]$ with the $\mathbf{s}_\rho$ as columns is then unitary, and we have

$$(\mathbf{S}^*\mathbf{W}\mathbf{S})_{11} = \mathbf{s}_1^*\mathbf{W}\mathbf{s}_1 = \alpha_1\mathbf{s}_1^*\mathbf{s}_1 = \alpha_1,$$

$$(\mathbf{S}^*\mathbf{W}\mathbf{S})_{\rho 1} = \mathbf{s}_\rho^*\mathbf{W}\mathbf{s}_1 = \alpha_1\mathbf{s}_\rho^*\mathbf{s}_1 = 0 \quad (\rho \neq 1),$$

$$(\mathbf{S}^*\mathbf{W}\mathbf{S})_{1\rho} = \mathbf{s}_1^*\mathbf{W}\mathbf{s}_\rho = (\bar{\alpha}_1)^{-1}\mathbf{s}_1^*\mathbf{W}^*\mathbf{W}\mathbf{s}_\rho = (\bar{\alpha}_1)^{-1}\mathbf{s}_1^*\mathbf{s}_\rho = 0 \quad (\rho \neq 1).$$

Hence the matrix $\mathbf{Q} = \mathbf{S}^*\mathbf{W}\mathbf{S}$ is of the form

$$\begin{bmatrix} \alpha_1 & 0 & \ldots & 0 \\ 0 & q_{22} & \cdots & q_{2r} \\ \ldots & \ldots & \ldots & \ldots \\ 0 & q_{r2} & \cdots & q_{rr} \end{bmatrix} = \begin{bmatrix} \alpha_1 & \mathbf{0}^* \\ \mathbf{0} & \mathbf{Q}_1 \end{bmatrix};$$

since $\mathbf{Q}$, being a product of unitary matrices, is itself unitary, it follows that the $(r-1)\times(r-1)$ matrix $\mathbf{Q}_1$ is also unitary. By the inductive hypothesis, there is an $(r-1)\times(r-1)$ unitary matrix $\mathbf{P}_1$ such that $\mathbf{P}_1^*\mathbf{Q}_1\mathbf{P}_1$ is diagonal; we border this to form the $r\times r$ unitary matrix

$$\mathbf{P}=\begin{bmatrix}1 & \mathbf{0}^* \\ \mathbf{0} & \mathbf{P}_1\end{bmatrix}.$$

Then $\mathbf{P}^*\mathbf{Q}\mathbf{P}$ is diagonal, i.e.

$$\mathbf{D}=\mathbf{P}^*\mathbf{S}^*\mathbf{W}\mathbf{S}\mathbf{P}=(\mathbf{S}\mathbf{P})^*\,\mathbf{W}(\mathbf{S}\mathbf{P})$$

is diagonal; since $\mathbf{T}=\mathbf{S}\mathbf{P}$ is unitary, the first result is proved. Finally, we have

$$\begin{aligned}\det(\mathbf{D}-\alpha\mathbf{I})&=\det[\mathbf{T}^*(\mathbf{W}-\alpha\mathbf{I})\,\mathbf{T}]\\&=(\det\mathbf{T}^{-1})\det(\mathbf{W}-\alpha\mathbf{I})\det\mathbf{T}\\&=\det(\mathbf{W}-\alpha\mathbf{I});\end{aligned}$$

thus $\mathbf{W}$ and $\mathbf{D}$ have the same eigenvalues and, since $\mathbf{D}$ is diagonal, its eigenvalues are just its diagonal elements.

(iii) We now return to the proof of Theorem 8·6·4. By Lemma 2, there is a unitary matrix $\mathbf{T}=[t_{\sigma\rho}]$ such that $\mathbf{T}^*\mathbf{W}\mathbf{T}$ is equal to a diagonal matrix

$$\begin{bmatrix}\alpha_1 & 0 & \dots & 0\\ 0 & \alpha_2 & \dots & 0\\ \dots & \dots & \dots & \dots\\ 0 & 0 & \dots & \alpha_r\end{bmatrix},$$

where, by Lemma 1, $|\alpha_\rho|=1$ $(1\leqslant\rho\leqslant r)$. We now write

$$y_\rho=\sum_{\sigma=1}^{r}t_{\sigma\rho}u_\sigma\quad(1\leqslant\rho\leqslant r);\qquad(5)$$

since $\mathbf{T}$ is unitary, the functions $y_1, y_2, \dots, y_r$ form an orthonormal set, and

$$u_\rho=\sum_{\sigma=1}^{r}\bar{t}_{\rho\sigma}y_\sigma\quad(1\leqslant\rho\leqslant r).\qquad(6)$$

By (2), (5) and (6), we then have

$$\begin{aligned}\mu K y_\rho &= \sum_{\sigma=1}^{r} t_{\sigma\rho}(\mu K u_\sigma)\\ &= \sum_{\sigma=1}^{r} t_{\sigma\rho} \sum_{\tau=1}^{r} w_{\tau\sigma} u_\tau\\ &= \sum_{\sigma=1}^{r} \sum_{\tau=1}^{r} t_{\sigma\rho} w_{\tau\sigma} \sum_{\nu=1}^{r} \bar{t}_{\tau\nu} y_\nu\\ &= \sum_{\nu=1}^{r}\left(\sum_{\sigma=1}^{r}\sum_{\tau=1}^{r} \bar{t}_{\tau\nu} w_{\tau\sigma} t_{\sigma\rho}\right) y_\nu\\ &= \alpha_\rho y_\rho \quad (1 \leqslant \rho \leqslant r). \end{aligned} \tag{7}$$

Since $|\alpha_\rho| = 1$, (7) is equivalent to

$$y_\rho = \bar{\alpha}_\rho \mu K y_\rho \quad (1 \leqslant \rho \leqslant r).$$

Thus y_ρ is a characteristic function of K, the corresponding characteristic value being $\bar{\alpha}_\rho \mu$. At the same time, since y_ρ is a linear combination of $u_1, u_2, \ldots, u_r$, it is a characteristic function of $K\bar{K}^* = K^*K$ with characteristic value μ^2; it follows that

$$\mu K^* y_\rho = \bar{\alpha}_\rho \mu^2 K^* K y_\rho = \bar{\alpha}_\rho y_\rho \quad (1 \leqslant \rho \leqslant r), \tag{8}$$

whence, by Theorem 8·2·2, $[y_\rho, \bar{\alpha}_\rho y_\rho]$ is a pair of singular functions of K belonging to the singular value μ, and y_ρ is a characteristic function of K^* belonging to the characteristic value $\alpha_\rho \mu$.

(iv) It follows from (5) and (6) that every pair of singular functions of K belonging to the singular value μ can be expressed in terms of the pairs $[y_\rho, \bar{\alpha}_\rho y_\rho]$. We can now form the functions y_ρ corresponding to the various singular values of K into a single orthonormal system (x_n), arranged so that the sequence (μ_n) formed by the corresponding singular values is non-decreasing; we denote the sequence formed by the corresponding characteristic values of K by (λ_n). We thus obtain a singular system of K, which can be written in the form

$$(x_n, x_n \operatorname{sgn} \lambda_n; |\lambda_n|).$$

We observe that x_n is both a characteristic function of K with characteristic value λ_n and a characteristic function of K^* with characteristic value $\bar{\lambda}_n$.

(v) We next prove that $(x_n; \lambda_n)$ is a characteristic system of K. By Theorem 8·3·2, Corollary 2 we have, for arbitrary $\mathfrak{L}^2$ functions x and y,

$$(Kx, y) = \sum_{n=1}^{\infty} \frac{(x, x_n \operatorname{sgn} \lambda_n)(x_n, y)}{|\lambda_n|}$$

$$= \sum_{n=1}^{\infty} \frac{(x, x_n)(x_n, y)}{\lambda_n}.$$

In particular,
$$(Kx, x_p) = \sum_{n=1}^{\infty} \frac{(x, x_n)(x_n, x_p)}{\lambda_n}$$

$$= \frac{(x, x_p)}{\lambda_p} \quad (p \geqslant 1). \tag{9}$$

Also, by Theorem 8·3·2 itself,

$$Kx = {}^{\circ}\sum_{n=1}^{\infty} (Kx, x_n) x_n$$

for an arbitrary $\mathfrak{L}^2$ function x.

Now suppose that x is a characteristic function of K with characteristic value λ. We then have

$$x = \lambda Kx = {}^{\circ}\sum_{n=1}^{\infty} (\lambda Kx, x_n) x_n = \sum_{n=1}^{\infty} (x, x_n) x_n. \tag{10}$$

On the other hand, by (9),

$$(x, x_n) = (\lambda Kx, x_n) = \frac{\lambda(x, x_n)}{\lambda_n},$$

whence $(x, x_n) = 0$ unless $\lambda = \lambda_n$. If $\lambda \neq \lambda_n$ for all n, it follows from (10) that $x = {}^{\circ} 0$, whence $x = \lambda Kx = 0$; thus λ cannot be a characteristic value. If $\lambda = \lambda_n$ for some n, (10) gives

$$x = {}^{\circ}\sum_{\lambda_n = \lambda} (x, x_n) x_n,$$

whence
$$x = \lambda Kx = \sum_{\lambda_n = \lambda} (x, x_n) \lambda K x_n = \sum_{\lambda_n = \lambda} (x, x_n) x_n,$$

so that x is a finite linear combination of members of the sequence (x_n). Thus every characteristic function of K is a linear combination of the (x_n); in other words, $(x_n; \lambda_n)$ is a characteristic system of K.

(vi) Let us now suppose that $(z_n; \lambda_n)$ is an arbitrary characteristic system of K. We have $z_n = \lambda_n K z_n$ and, by Theorem 8·6·3, we also have $z_n = \bar{\lambda}_n K^* z_n$. Writing $\mu_n = |\lambda_n|$, we obtain

$$\mu_n K^* z_n = |\lambda_n| K^* z_n = (\bar{\lambda}_n \operatorname{sgn} \lambda_n) K^* z_n = (\operatorname{sgn} \lambda_n) z_n.$$

Putting $y_n = (\operatorname{sgn} \lambda_n) z_n$, we also have

$$\mu_n K y_n = (|\lambda_n| \operatorname{sgn} \lambda_n) K z_n = \lambda_n K z_n = z_n.$$

These equations together show that, for every value of n, $[z_n, (\operatorname{sgn} \lambda_n) z_n]$ is a pair of singular functions of K belonging to the singular value $\mu_n = |\lambda_n|$.

Since the full system (z_n) of characteristic functions of K is linearly equivalent to the particular system (x_n) constructed in parts (i)–(v) of this proof, and $(x_n, x_n \operatorname{sgn} \lambda_n; |\lambda_n|)$ is a singular system of K, it follows at once that $(z_n, z_n \operatorname{sgn} \lambda_n; |\lambda_n|)$ is also a singular system of K. The proof of the theorem is therefore complete.

THEOREM 8·6·5. *Let K be a normal $\mathfrak{L}^2$ kernel, and let $(x_n; \lambda_n)$ be a characteristic system of K. Then*

$$K = {}^\circ\sum_{n=1}^{\infty} \frac{x_n \otimes x_n}{\lambda_n} \quad (\mathfrak{L}^2), \tag{11}$$

$$\| K \|^2 = \sum_{n=1}^{\infty} \frac{1}{|\lambda_n|^2}; \tag{12}$$

if x is an arbitrary $\mathfrak{L}^2$ function and $y = Kx$, then

$$y = {}^\circ\sum_{n=1}^{\infty} (y, x_n) x_n = \sum_{n=1}^{\infty} \frac{(x, x_n)}{\lambda_n} x_n, \tag{13}$$

the series being relatively uniformly absolutely convergent; finally, if x and y are arbitrary $\mathfrak{L}^2$ functions, then

$$(Kx, y) = \sum_{n=1}^{\infty} \frac{(x, x_n)(x_n, y)}{\lambda_n}. \tag{14}$$

By Theorem 8·6·4, $(x_n, x_n \operatorname{sgn} \lambda_n; |\lambda_n|)$ is a singular system of K. All the equations (11)–(14) then follow at once from Theorems 8·3·2 and 8·3·3.

We see from Theorem 8·6·5 that normal kernels behave just like Hermitian kernels, with the single exception that their characteristic values are not necessarily real. The main results of §§ 7·4–7·6 can be extended at once to normal kernels; in fact, the only serious change needed in the proofs occurs in the case $p=2$ of Theorem 7·6·1, where Dini's theorem has to be applied to $KK^*(s,s)$ instead of to $K^2(s,s)$.

We next prove an analogue of Theorem 7·8·1.

THEOREM 8·6·6. *Let $(x_n; \lambda_n)$ be a characteristic system of the normal $\mathfrak{L}^2$ kernel K; write $\kappa_n = \lambda_n^{-1}$ $(n \geqslant 1)$. Then*

$$|\kappa_n| = \sup\{|(Kx,x)| : \|x\|=1, (x,x_1)=\ldots=(x,x_{n-1})=0\}$$
$$= \sup\{\|Kx\| : \|x\|=1, (x,x_1)=\ldots=(x,x_{n-1})=0\}.$$

By (14), if $\|x\|=1$ and $(x,x_1)=\ldots=(x,x_{n-1})=0$,

$$(Kx,x) = \sum_{\nu=n}^{\infty} \frac{(x,x_\nu)(x_\nu,x)}{\lambda_\nu}$$
$$= \sum_{\nu=n}^{\infty} \kappa_\nu |(x,x_\nu)|^2,$$

so that
$$|(Kx,x)| \leqslant \sum_{\nu=n}^{\infty} |\kappa_\nu| \cdot |(x,x_\nu)|^2$$
$$\leqslant |\kappa_n| \sum_{\nu=n}^{\infty} |(x,x_\nu)|^2$$
$$\leqslant |\kappa_n| \cdot \|x\|^2 = |\kappa_n|.$$

On the other hand, if we take $x = x_n$, we obtain

$$(Kx,x) = \kappa_n \|x_n\|^2 = \kappa_n;$$

the first equation of the theorem follows at once. To obtain the second equation, we remark that, by (13),

$$Kx = {}^\circ\sum_{\nu=n}^{\infty} \kappa_\nu (x,x_\nu) x_\nu,$$

whence
$$\|Kx\|^2 = \sum_{\nu=n}^{\infty} |\kappa_\nu|^2 |(x,x_\nu)|^2,$$

and we then argue as before.

Analogues of Theorems 7·9·1–7·9·3, concerned with the resolvent of the kernel and the solution of the linear integral equation of the second kind, can be proved without difficulty; details are left to the reader.

Before proving the next major theorem, we require the following auxiliary result, which has some independent interest.

THEOREM 8·6·7. *If the $\mathfrak{L}^2$ kernel K is almost normal, i.e.* $KK^* =^\circ K^*K$, *then K is equivalent to a normal kernel.*

Let K be an almost normal $\mathfrak{L}^2$ kernel. By following the proofs of Theorems 8·6·3 and 8·6·4, replacing equality by equivalence when necessary, we can show that K has a singular system $(x_n, y_n; |\lambda_n|)$ such that $(x_n; \lambda_n)$ is a characteristic system for K, $(y_n; \bar{\lambda}_n)$ is a characteristic system for K^*, and $y_n =^\circ (\operatorname{sgn} \lambda_n) x_n$ for all n. To the orthonormal system (x_n) we can adjoin (§ 4·1) a finite or enumerable sequence (z_n) of $\mathfrak{L}^2$ functions in such a way as to obtain a complete orthonormal system (e_n). Since $\overline{K(s,u)}$ and $\overline{K(t,u)}$ are $\mathfrak{L}^2$ functions of u for each value of s and t respectively, the generalized Parseval equation (§ 4·3) gives

$$KK^*(s,t) = \int K(s,u)\,\overline{K(t,u)}\,du$$

$$= \sum_{n=1}^{\infty} \left\{\int K(s,u)\, e_n(u)\, du\right\}\left\{\int \overline{K(t,u)}\,\overline{e_n(u)}\, du\right\}$$

$$= \sum_{n=1}^{\infty} Ke_n(s)\,.\,\overline{Ke_n(t)}$$

$$= \sum_{n=1}^{\infty} Kx_n(s)\,.\,\overline{Kx_n(t)} + \sum_{n=1}^{\infty} Kz_n(s)\,.\,\overline{Kz_n(t)}.$$

Similarly,

$$K^*K(s,t) = \sum_{n=1}^{\infty} K^*x_n(s)\,.\,\overline{K^*x_n(t)} + \sum_{n=1}^{\infty} K^*z_n(s)\,.\,\overline{K^*z_n(t)}.$$

Now $\lambda_n Kx_n = x_n$, and $\bar{\lambda}_n K^* x_n = y_n \overline{\operatorname{sgn} \lambda_n} =^\circ x_n$; hence $\lambda_n Kx_n =^\circ \bar{\lambda}_n K^* x_n$. Also, since each z_n is orthogonal to all x_n (and therefore to all y_n), we have $Kz_n =^\circ 0$, $K^*z_n =^\circ 0$, whence $Kz_n =^\circ K^*z_n$. We now modify the kernel $K(s,t)$, obtaining an equivalent kernel $K_0(s,t)$, as follows: if, for a particular value of s, we have

$$\lambda_n \int K(s,u)\, x_n(u)\, du \neq \bar{\lambda}_n \int \overline{K(u,s)}\, x_n(u)\, du$$

for some n, or either of the relations

$$\int K(s,u)\, z_n(u)\, du \neq 0, \quad \int \overline{K(u,s)}\, z_n(u)\, du \neq 0,$$

holds for some n, we put $K_0(s,u)=0$ for all u and $K_0(u,s)=0$ for all u; otherwise we put $K_0(s,t)=K(s,t)$. We shall then have

$$\begin{aligned} K_0 K_0^*(s,t) &= \sum_{n=1}^{\infty} K_0 x_n(s) . \overline{K_0 x_n(t)} \\ &= \sum_{n=1}^{\infty} \frac{\lambda_n K_0 x_n(s) . \overline{\lambda_n K_0 x_n(t)}}{\lambda_n \overline{\lambda}_n} \\ &= \sum_{n=1}^{\infty} \frac{\overline{\lambda}_n K_0^* x_n(s) . \overline{\lambda_n K_0^* x_n(t)}}{\lambda_n \overline{\lambda}_n} \\ &= \sum_{n=1}^{\infty} K_0^* x_n(s) . \overline{K_0^* x_n(t)} \\ &= K_0^* K_0(s,t). \end{aligned}$$

Thus the kernel $K_0(s,t)$, which is equivalent to $K(s,t)$, is a normal kernel.

The above proof is deeper than one might have expected, but no more elementary proof is known.

We shall now show that the normal kernels form the widest class of $\mathfrak{L}^2$ kernels that can be expressed in terms of their characteristic functions and characteristic values in the way described in Theorem 8·6·5. The following theorem, which asserts an even stronger result, was first proved by Goldfain (1946).

THEOREM 8·6·8. *Let K be an $\mathfrak{L}^2$ kernel, and let (λ_n) be its sequence of characteristic values, each repeated a number of times equal to its rank. Then*

$$\sum_{n=1}^{\infty} \frac{1}{|\lambda_n|^2} \leqslant \| K \|^2, \tag{15}$$

and K is equivalent to a normal kernel if and only if equality holds in (15).

Let (x_n) be a linearly independent sequence of characteristic functions of K, normalized to make $\| x_n \| = 1$, and such that

$$x_n = \lambda_n K x_n \quad (n \geqslant 1).$$

For convenience, we write $\kappa_n = \lambda_n^{-1}$, so that

$$Kx_n = \kappa_n x_n \quad (n \geqslant 1).$$

By arguments of a familiar type (see, for example, p. 120), we see that the sequence (x_n) is in fact linearly independent with respect to equivalence. It then follows from the proof of Theorem 4·4·1 that there is an orthonormal system (y_n) such that, for each n, y_n is a linear combination of $x_1, x_2, \ldots, x_n$ and x_n is a linear combination of $y_1, y_2, \ldots, y_n$. Suppose that

$$y_n = \sum_{\nu=1}^{n} \alpha_{n\nu} x_\nu \quad (n \geqslant 1).$$

Then
$$Ky_n = \sum_{\nu=1}^{n} \alpha_{n\nu} \kappa_\nu x_\nu = \sum_{\nu=1}^{n} \beta_{n\nu} y_\nu,$$

say. From the construction of the sequence (y_n) it follows at once that $\beta_{nn} = \kappa_n$; hence

$$Ky_n = \sum_{\nu=1}^{n-1} \beta_{n\nu} y_\nu + \kappa_n y_n \quad (n \geqslant 1),$$

i.e.
$$\int K(s,t)\, y_n(t)\, dt = \sum_{\nu=1}^{n-1} \beta_{n\nu} y_\nu(s) + \kappa_n y_n(s) \quad (n \geqslant 1).$$

Since $K(s,t)$ is an $\mathfrak{L}^2$ function of t for fixed s, Bessel's inequality gives
$$\sum_{n=1}^{N} \left| \sum_{\nu=1}^{n-1} \beta_{n\nu} y_\nu(s) + \kappa_n y_n(s) \right|^2 \leqslant \int |K(s,t)|^2\, dt \quad (N \geqslant 1).$$

Integrating with respect to s, and using the fact that (y_n) is orthonormal, we obtain

$$\sum_{n=1}^{N} \sum_{\nu=1}^{n-1} |\beta_{n\nu}|^2 + \sum_{n=1}^{N} |\kappa_n|^2 \leqslant \| K \|^2 \quad (N \geqslant 1),$$

whence
$$\sum_{n=1}^{\infty} \sum_{\nu=1}^{n-1} |\beta_{n\nu}|^2 + \sum_{n=1}^{\infty} |\kappa_n|^2 \leqslant \| K \|^2. \tag{16}$$

The first part of the theorem follows at once.

Suppose now that
$$\sum_{n=1}^{\infty} |\kappa_n|^2 = \| K \|^2. \tag{17}$$

To reconcile (16) and (17), we must have

$$\beta_{n\nu} = 0 \quad (1 \leqslant \nu < n),$$

whence
$$Ky_n = \kappa_n y_n \quad (n \geqslant 1). \tag{18}$$

Thus (y_n) is an orthonormal system of characteristic functions of K; furthermore, every characteristic function of K must be a finite linear combination of the (y_n). We now expand the expression

$$\left\| K - \sum_{\nu=1}^{n} \kappa_\nu (y_\nu \otimes y_\nu) \right\|^2.$$

Using (18), we obtain

$$\begin{aligned}
\iint \left| K(s,t) - \sum_{\nu=1}^{n} \kappa_\nu y_\nu(s) \overline{y_\nu(t)} \right|^2 ds\,dt
&= \iint \left[K(s,t) - \sum_{\nu=1}^{n} \kappa_\nu y_\nu(s) \overline{y_\nu(t)} \right] \\
&\quad \times \left[\overline{K(s,t)} - \sum_{\rho=1}^{n} \bar{\kappa}_\rho \overline{y_\rho(s)} y_\rho(t) \right] ds\,dt \\
&= \| K \|^2 - \sum_{\rho=1}^{n} \kappa_\rho \bar{\kappa}_\rho \| y_\rho \|^2 - \sum_{\nu=1}^{n} \kappa_\nu \bar{\kappa}_\nu \| y_\nu \|^2 \\
&\quad + \sum_{\nu,\rho=1}^{n} \kappa_\nu \bar{\kappa}_\rho (y_\nu, y_\rho)(y_\rho, y_\nu) \\
&= \| K \|^2 - \sum_{\nu=1}^{n} | \kappa_\nu |^2 \\
&\to 0 \quad (n \to \infty),
\end{aligned}$$

i.e.

$$\begin{aligned}
K &= {}^{\circ}\sum_{n=1}^{\infty} \kappa_n (y_n \otimes y_n) \quad (\mathfrak{L}^2) \\
&= {}^{\circ}\sum_{n=1}^{\infty} \frac{y_n \otimes y_n}{\lambda_n} \quad (\mathfrak{L}^2).
\end{aligned}$$

It follows at once that

$$K^* = {}^{\circ}\sum_{n=1}^{\infty} \frac{y_n \otimes y_n}{\bar{\lambda}_n} \quad (\mathfrak{L}^2),$$

whence

$$KK^* = {}^{\circ}\sum_{n=1}^{\infty} \frac{y_n \otimes y_n}{|\lambda_n|^2} = {}^{\circ} K^*K \quad (\mathfrak{L}^2).$$

Thus K is almost normal; by Theorem 8·6·7, K is equivalent to a normal kernel.

Conversely, suppose that K is equivalent to a normal kernel, and therefore almost normal. From the remarks at the beginning of the proof of Theorem 8·6·7 and from Theorem 8·6·5 it then follows at once that

$$\sum_{n=1}^{\infty} \frac{1}{|\lambda_n|^2} = \| K \|^2.$$

8·7. Linear integral equations of the first kind. In this section we shall show how the theory of singular functions and singular values can be used to obtain a criterion of solubility for linear equations of the first kind.

THEOREM 8·7·1.† *Let $(u_n, v_n; \mu_n)$ be a singular system of the $\mathfrak{L}^2$ kernel $K(s,t)$, and let $y(s)$ be a given $\mathfrak{L}^2$ function. Then the equation*

$$y(s) = {}^\circ \int K(s,t)\, x(t)\, dt \tag{1}$$

has an $\mathfrak{L}^2$ solution $x(t)$ if and only if (a)

$$\sum_{n=1}^{\infty} \mu_n^2 \,|\, (y, u_n) \,|^2 < +\infty,$$

and (b) $(y,u)=0$ for every $\mathfrak{L}^2$ function u such that $K^*u = {}^\circ 0$.

Suppose first that $x(t)$ is an $\mathfrak{L}^2$ solution of (1). Then

$$(y, u_n) = (Kx, u_n) = (x, K^*u_n) = \frac{(x, v_n)}{\mu_n},$$

whence $$(x, v_n) = \mu_n (y, u_n) \quad (n \geqslant 1).$$

It now follows from Bessel's inequality, applied to x and the orthonormal system (v_n), that

$$\sum_{n=1}^{\infty} \mu_n^2 \,|\, (y, u_n) \,|^2 \leqslant \| x \|^2 < +\infty;$$

this is condition (*a*). Furthermore, if u is an $\mathfrak{L}^2$ solution of the equation $K^*u = {}^\circ 0$, we have

$$(y, u) = (Kx, u) = (x, K^*u) = 0;$$

thus condition (*b*) holds.

Conversely, suppose that conditions (*a*) and (*b*) are both satisfied. Then, by the Riesz–Fischer theorem (Theorem 4·3·2), there is an $\mathfrak{L}^2$ function x such that

$$(x, v_n) = \mu_n (y, u_n) \quad (n \geqslant 1).$$

Hence, by Theorem 8·3·2,

$$Kx = {}^\circ \sum_{n=1}^{\infty} \frac{(x, v_n)}{\mu_n} u_n = \sum_{n=1}^{\infty} (y, u_n)\, u_n, \tag{2}$$

† Picard (1910).

the series being relatively uniformly absolutely convergent. Let us now write

$$z = y - \sum_{n=1}^{\infty} (y, u_n)\, u_n; \tag{3}$$

a routine calculation then shows that

$$\| z \|^2 = \| y \|^2 - \sum_{n=1}^{\infty} | (y, u_n) |^2. \tag{4}$$

Also, we have

$$(z, u_n) = (y, u_n) - (y, u_n) = 0 \quad (n \geqslant 1),$$

so that, by Theorem 8·3·2,

$$K^* z = {}^\circ \sum_{n=1}^{\infty} \frac{(z, u_n)}{\mu_n} v_n = 0.$$

By condition (*b*) it follows that $(y, z) = 0$; substituting for z from (3), we obtain

$$0 = (y, z) = \| y \|^2 - \sum_{n=1}^{\infty} | (y, u_n) |^2,$$

whence, by (4), $\| z \| = 0$, $z = {}^\circ 0$. Consequently, by (3) and (2),

$$y = {}^\circ \sum_{n=1}^{\infty} (y, u_n)\, u_n = {}^\circ Kx,$$

i.e. x is an $\mathfrak{L}^2$ solution of (1).

When $K(s, t)$ is continuous, the above result assumes the following form.

THEOREM 8·7·2. *Let* $(u_n, v_n; \mu_n)$ *be a singular system of the continuous kernel* $K(s, t)$, *and let* $y(s)$ *be a given continuous function. Then the equation*

$$y(s) = \int K(s, t)\, x(t)\, dt$$

has an $\mathfrak{L}^2$ *solution* $x(t)$ *if and only if* (*a*)

$$\sum_{n=1}^{\infty} \mu_n^2 | (y, u_n) |^2 < +\infty,$$

and (*b*) $(y, u) = 0$ *for every* $\mathfrak{L}^2$ *function* u *such that* $K^* u = 0$.

We note that, even when $K(s, t)$ and $y(s)$ are continuous, the solution $x(t)$ is not necessarily continuous; examples can easily be constructed to illustrate this fact.

We note also that the $\mathfrak{L}^2$ solution x of (1) is unique (up to equivalence) if and only if $Kx = {}^\circ 0$ implies $x = {}^\circ 0$; by Theorem 8·3·1, this amounts to saying that $(x, v_n) = 0$ for all n implies $x = {}^\circ 0$; in other words, (v_n) is a complete orthonormal system.

Finally, we remark that equation (1) can never possess an $\mathfrak{L}^2$ solution for every given $\mathfrak{L}^2$ function $y(s)$. If this could happen, conditions (*a*) and (*b*) would hold for all y; this implies in the first place that $K^*u = {}^\circ 0$ implies $u = {}^\circ 0$ or, in other words, that (u_n) is a complete orthonormal system. The sequence (μ_n) is therefore genuinely infinite; since, by the Riesz–Fischer theorem, $((y, u_n))$ may be any sequence (η_n) of complex numbers for which $\Sigma |\eta_n|^2 < +\infty$, we should have, writing $\alpha_n = |\eta_n|^2$,

$$\sum_{n=1}^{\infty} \alpha_n \mu_n^2 < +\infty$$

for every sequence (α_n) of positive numbers such that $\Sigma \alpha_n < +\infty$. Since $\mu_n \to \infty$ $(n \to \infty)$, we have a contradiction.

The above indications are sufficient to show that linear integral equations of the first kind behave very differently from those of the second kind.

BIBLIOGRAPHY

BIRKHOFF, G. (1948). *Lattice theory*. Revised edition. American Mathematical Society Colloquium Publications, no. 25. New York.

BÔCHER, M. (1909). *An introduction to the study of integral equations*. Cambridge Tracts in Mathematics and Mathematical Physics, no. 10. Cambridge University Press.

BÜCKNER, H. (1952). *Die praktische Behandlung von Integralgleichungen*. Ergebnisse der angewandten Mathematik, no. 1. Berlin, Göttingen, Heidelberg.

BURKILL, J. C. (1951). *The Lebesgue integral*. Cambridge Tracts in Mathematics and Mathematical Physics, no. 40. Cambridge University Press.

CARLEMAN, T. (1921). Zur Theorie der linearen Integralgleichungen. *Math. Z.* **9**, 196–217.

COURANT, R. & HILBERT, D. (1953). *Methods of mathematical physics*, vol. 1. New York.

DINI, U. (1878). *Fondamenti per la teoria delle funzioni di variabili reali*. Pisa.

FAN, KY (1949). On a theorem of Weyl concerning eigenvalues of linear transformations. I. *Proc. Nat. Acad. Sci. U.S.A.* **35**, 652–5.

FISCHER, E. (1907). Sur la convergence en moyenne. *C.R. Acad. Sci., Paris*, **144**, 1022–4.

FREDHOLM, I. (1903). Sur une classe d'équations fonctionnelles. *Acta math.*, Stockh., **27**, 365–90.

GOLDFAIN, I. A. (1946). Sur une classe d'équations intégrales linéaires. *Uchenye Zapiski Moskov. Gos. Univ.* 100, Matematika, tom 1, 104–12.

GOURSAT, E. (1927). *Cours d'analyse mathématique*, 4th ed., vol. 3. Paris.

GRAM, J. P. (1883). Ueber die Entwickelung reeller Functionen in Reihen mittelst der Methode der kleinsten Quadraten. *J. reine angew. Math.* **94**, 41–73.

HADAMARD, J. (1893). Résolution d'une question relative aux déterminants. *Bull. Sci. Math.* (2), **17**, 240–6.

HARDY, G. H., LITTLEWOOD, J. E. & PÓLYA, G. (1934). *Inequalities*. Cambridge University Press.

HELLINGER, E. & TOEPLITZ, O. (1927). Integralgleichungen und Gleichungen mit unendlichvielen Unbekannten. *Encyklopädie der Mathematischen Wissenschaften*, II.C.13. Band II, 3. Teil, 2. Hälfte, 1335–1601. Leipzig.

HILBERT, D. (1904). Grundzüge einer allgemeinen Theorie der linearen Integralgleichungen. Erste Mitteilung. *Nachr. Ges. Wiss. Göttingen*, 1904, 49–91.

HOBSON, E. W. (1927). *The theory of functions of a real variable and the theory of Fourier series*, 3rd ed., vol. 1. Cambridge University Press.

Kaczmarz, S. & Steinhaus, H. (1935). *Theorie der Orthogonalreihen.* Monografje Matematyczne, no. 6. Warszawa.

Kneser, A. (1906). Ein Beitrag zur Theorie der Integralgleichungen. *R.C. mat. Palermo*, **22**, 233–40.

Le Roux, J. (1895). Sur les intégrales des équations linéaires aux dérivées partielles du second ordre à deux variables indépendantes. *Ann. Sci. Ec. Norm. Sup.* (3), **12**, 227–316.

Lewis, D. C. (1950). Comments on the classical theory of integral equations. *J. Wash. Acad. Sci.* **40**, 65–71.

Liouville, J. (1837). Second mémoire sur le développement des fonctions en séries dont les divers termes sont assujettis à satisfaire à une même équation différentielle du second ordre contenant un paramètre variable. *J. Math. pures appl.* (1), **2**, 16–35.

Lovitt, W. V. (1924). *Linear integral equations.* New York.

Mercer, J. (1909). Functions of positive and negative type, and their connection with the theory of integral equations. *Phil. Trans.* A, **209**, 415–46.

Michal, A. D. & Martin, R. S. (1934). Some expansions in vector space. *J. Math. Pures Appl.* (9), **13**, 69–91.

Mollerup, J. (1924). Sur l'itération d'une fonction par un noyau donné. *R.C. mat. Palermo*, **47**, 375–95.

Moore, E. H. (1910). Introduction to a form of general analysis. *New Haven Mathematical Colloquium*, pp. 1–150. New Haven.

Neumann, C. (1877). *Untersuchungen über das logarithmische und Newtonsche Potential.* Leipzig.

Picard, E. (1910). Sur un théorème générale relatif aux équations intégrales de première espèce et sur quelques problèmes de physique mathématique. *R.C. mat. Palermo*, **29**, 79–97.

Plemelj, J. (1904). Zur Theorie der Fredholmschen Funktionalgleichungen. *Mh. Math. Phys.* **15**, 93–128.

Radon, J. (1919). Über lineare Funktionaltransformationen und Funktionalgleichungen. *S.B. Akad. Wiss. Wien*, **128** (Abteilung II*a*), 1083–121.

Riesz, F. (1907*a*). Sur les systèmes orthogonaux de fonctions. *C.R. Acad. Sci., Paris*, **144**, 615–19.

Riesz, F. (1907*b*). Über orthogonale Funktionensysteme. *Nachr. Ges. Wiss. Göttingen*, 1907, 116–22.

Riesz, F. (1910). Untersuchungen über Systeme integrierbarer Funktionen. *Math. Ann.* **69**, 449–97.

Ruston, A. F. (1951). On the Fredholm theory of integral equations for operators belonging to the trace class of a general Banach space. *Proc. Lond. math. Soc.* (2), **53**, 109–24.

Salem, R. (1954). On a problem of Smithies. *Proc. Acad. Sci. Amst.* A, 57 = *Indagationes Math.* **16**, 403–7.

Schmidt, E. (1907*a*). Zur Theorie der linearen und nichtlinearen Integralgleichungen. Erster Teil. Entwicklung willkürlicher Funktionen nach Systemen vorgeschriebener. *Math. Ann.* **63**, 433–76.

SCHMIDT, E. (1907*b*). Zur Theorie der linearen und nichtlinearen Integralgleichungen. Zweite Abhandlung. Auflösung der allgemeinen linearen Integralgleichung. *Math. Ann.* **64**, 161–74.

SMITHIES, F. (1935). On the theory of linear integral equations. *Proc. Camb. Phil. Soc.* **31**, 76–84.

SMITHIES, F. (1937). The eigen-values and singular values of integral equations. *Proc. Lond. math. Soc.* (2), **43**, 255–79.

SMITHIES, F. (1941). The Fredholm theory of integral equations. *Duke Math. J.* **8**, 107–30.

TITCHMARSH, E. C. (1932). *The theory of functions.* Oxford.

VERGERIO, A. (1919). Sulle equazioni integrali di prima specie a nucleo non simmetrico. *R.C. mat. Palermo*, **42**, 285–302.

VOLTERRA, V. (1897). Sopra alcune questioni di inversione di integrali definiti. *Ann. di. Mat.* (2), **25**, 139–78.

WEYL, H. (1909). Über die Konvergenz von Reihen, die nach Orthogonalfunktionen fortschreiten. *Math. Ann.* **67**, 225–45.

ZAANEN, A. C. (1953). *Linear analysis.* Bibliotheca mathematica, vol. 2. Amsterdam: Groningen.

INDEX

INDEX OF NOTATIONS